AF768325

GLOBALISATION AND SUSTAINABLE DEVELOPMENT: QUEST FOR A NEW PARADIGM

GLOBALISATION AND SUSTAINABLE DEVELOPMENT: QUEST FOR A NEW PARADIGM

Edited by:

Sheila Rai and
Dhirendra Vajpeyi

Delve
PUBLISHING

www.delvepublishing.com

Globalisation and Sustainable Development: Quest For A New Paradigm

Sheila Raj and Dhirendra Vajpeyi

Delve Publishing
4164 Lakeshore Road
Burlington, ON L7L 1A4
Canada
www.delvepublishing.com
Email: orders@arclereducation.com

© 2025 Delve Publishing

ISBN: 978-1-77469-008-6 (Hardcover)

This book contains information obtained from highly regarded resources. Reprinted material sources are indicated and copyright remains with the original owners. Copyright for images and other graphics remains with the original owners as indicated. A Wide variety of references are listed. Reasonable efforts have been made to publish reliable data. Authors or Editors or Publishers are not responsible for the accuracy of the information in the published chapters or consequences of their use. The publisher assumes no responsibility for any damage or grievance to the persons or property arising out of the use of any materials, instructions, methods or thoughts in the book. The authors or editors and the publisher have attempted to trace the copyright holders of all material reproduced in this publication and apologize to copyright holders if permission has not been obtained. If any copyright holder has not been acknowledged, please write to us so we may rectify.

Notice: Registered trademark of products or corporate names are used only for explanation and identification without intent of infringement.

Delve Publishing publishes wide variety of books and eBooks. For more information about Delve Publishing and its products, visit our website at www.delvepublishing.com.

ABOUT THE EDITORS

Sheila Rai - Associate Professor of Political Science, University of Rajasthan, Jaipur and presently serving as Principal, St. Xavier's College, Jaipur has more than 35 years of teaching and research experience. She has also served as faculty for the University of Guelph, Canada, SANN International University, Nepal and for students from various European countries. With a PhD in Gandhian Philosophy she has supervised many doctoral theses and also published several books and research papers on subjects pertaining to Gandhian thought, Peace Studies and Contemporary Politics both in English and Hindi. Her other areas of specific interest are Indian Government and Politics, Gender, Indigenization and Development studies. Rai has completed Major Research Projects on 'Evolution of Article 370: An Analysis', 'Caste, Party and Ticket Distribution in Rajasthan, 1952 – 2004', 'Gandhian Paradigm: Quintessence of Real Development'. She is the Editor of 'Rajyashti- Journal of Social Sciences, (biannual journal) of Rajasthan Political Science Association, also has 6 books and 45 national and international research publications to her credit. Some of her measure publications are related to the Gandhian Way: Its Contemporary Relevance (in Hindi); Social, Economic and Political Justice: Constitutional Implications and Practical Dimensions (in Hindi); Role of Party Politics and Caste in the Selection of Candidate (in Hindi); Indian Political System; Gandhian View of State, Individual and Society; Countering the Cult: An Assessment of the Contemporary Women led Movements in India; Article 35A & its implications; Water Disputes in South Asia in Water Resource Conflicts and International Security; Fragmented Responses towards Global Governance; Empowerment of Women through Education; Rashtriya Swasthya Beema Yojna: Panacea for the Poor; Corporate Social Responsibility in India; Globalization, Sustainable Development and Indigenous Technology. At present Rai is the President of Indian Society of Gandhian Studies, General Secretary National Social Science Association and Chair- RC 35, IPSA.

Dhirendra Vajpeyi - is Emeritus Professor of Political Science and Asian Studies at the University of Northern Iowa. He has authored, coauthored and edited seventeen books, numerous chapters and articles on international security, human rights and environmental policymaking. Noteworthy books include Corporate Social Responsibility and Sustainable Development in Emerging Economies; Climate Change Sustainable Development and Human Security; Environmental Policymaking in the Third world; Technology and Development; Local Government and Politics in the Third World; Indira Gandhi's India; Deforestation: A Comparative Analysis; Civil- Military Relations: Nation-building and National Identity; Local Democracy and Politics in South Asia; Modernizing China; Water Resource Management : A Comparative Perspective; Law, Politics and Society in India; Politics, Technology and Bureaucracy in South Asia; Globalization, Governance and Technology: Challenges and Alternatives, Water Resource Conflicts and International Security: A Global Perspective, Civil -military Relations in Developing Countries; Globalization Reappraised False Oracle or a Talisman. Vajpeyi has lectured at universities in Moscow, St. Petersburg (Russia), People's Republic of China, Nigeria, Chile, and Klagenfurt (Austria). He was a senior Research Fellow at Hoover Institute at Stanford University (California). Currently he is emeritus Co-Chair of IPSA.

TABLE OF CONTENTS

LIST OF CONTRIBUTORS

Ayesha Akbar is a scholar of National School of Public Policy, Pakistan.

Gupta Ashok Kumar with a Ph.D. in Geography is an Associate Professor in BSR Government Arts P.G. College, Alwar. Agriculture, Population and Environment are his special areas of interest and published research papers on related subjects.

Kacker Snehil is an Assistant Professor in the Department of Political Science at Lady Shri Ram College for Women in New Delhi. She has a Ph.D. in Politics of Climate Change and her areas of interest include Indian Government and Politics, Politics of Climate Change, Peace and Conflict Studies. She has presented papers at various National and International Conferences.

Khawaja Sarfraz received his Ph.D. from University of Missouri, USA and was a faculty member at the University of Wisconsin. He was awarded Curator's and Frank Fellowship and served as a visiting Fellow at Harvard University. He has authored 12 books and papers in impact factor journals. He is a frequent lecturer at civil service institutes and has been the consultant to The World Bank (WB), Asian Development Bank (ADB) and many international agencies. He has earned an illustrious career of working in civil services training institute, Ministry of Education, and UN-organizations. He has served as Chief Instructor Civil Services Academy at Lahore.

Kumari Radha earned her Ph.D. in Gandhian peace movements from University of Rajasthan. She is an Assistant Professor in Department of Political Science at Mata Sundri College in New Delhi and her area of interest is Political Theory and Gandhi. She has published in several scholarly journals on topics like Peace, Gandhi, Nation State and Violence.

Maurya Suman is MA gold medalist has a Ph.D. in Political Science and is working as an Assistant Professor in the Department of Political Science and Associate Dean, Students' welfare in University of Rajasthan, Jaipur. Maurya has received M N Roy Award and CSR Award. She has published 15 research papers in reputable academic journals. Her areas of interest are International relations, human rights and gender issues.

Sharma Meeta has her Ph.D. in Political Science with emphasis on Public Private Partnership and waste management strategies in Jaipur Municipal Corporation in partnership with private sector. She is presently an Assistant Professor of Department of Political Science at St. Xavier's College, Jaipur.

Sharma Preeti obtained her Ph.D. from the Department of Political Science, University of Rajasthan on the Role of Governor and State Politics in Rajasthan. She is presently an Assistant Professor of Political Science in St. Xavier's College, Jaipur. She has contributed 15 research papers in various scholarly journals and edited books. Her areas of interest are Political Theory and International Relations.

Shekhawat Vishwamaitri is an assistant professor of Geography in SMM Govt. Girls College, Bhilwara, Rajasthan. She had been a Research Scholar in the Department of Geography in University of Rajasthan. Her area of interest is Urban Geography.

Silvia Pellegrino is a graduate of the University of Northern Iowa. She holds BAs in Political Science and Global Studies with a focus on Peace, Conflict and Human Rights and the geographic areas of Africa and the Middle East. As a panelist at the 2019 Iowa Human Rights Conference, she was awarded the Weston Essay Prize for her paper "Can People Survive on Carrots and Sticks? Enhancing quality of life through financial tools of foreign policy following Sen's Capabilities Approach."

Vadlakonda Satyanarayana is Ph.D. from Kakatiya University. He has numerous articles to his credit published in both national and international peer reviewed journals. His specializations include Political Thought and Indian Politics. He has presented many papers both in national and international conferences. Currently he is the faculty of Political Science in the department of Political Science in Kakatiya University, Warangal, Telangana State, India.

Warby Brian earned his Ph.D. from the University of South Carolina and now teaches Political Science at the University of Northern Iowa. His doctoral dissertation and subsequent book examined the impact of political stability on the microfinance industry. He has also published research articles and book chapters that examine the role of gender and the effects of political institutions on economic development and poverty alleviation.

LIST OF FIGURES

LIST OF TABLES

INTRODUCTION

The summer of 2016 was memorable for several political and economic developments across the world. While unexpected Brexit threatened the European Union destabilizing Europe and creating security anxieties beyond Europe, the American President elect threatened to construct a wall between US and Mexico with the proclamation of bringing back jobs to the USA. With the rise of protectionist debate, ascendance of religious fundamentalism and ethnocentrism, proxy conflict in the Middle East, increased global terrorism and emergence of neo-Nazism in Europe, it is obvious that serious existentialist questions have emerged about the direction and the nature of economic and political policies. Several questions are being asked: (1) What is it really about? Immigration, economic inequality, xenophobia, state sovereignty, corrupt political elites, class war and neoliberal revolution; (2) What is to be done? (3) And how is it to be done? Many, both on left and right, with varying degrees blame globalization for the 'rise of economic dislocation'.

The collapse of Soviet Union in 1990 witnessed the end of Communist domination and the triumphal emergence of the Washington consensus inspired model of economic growth. The globalization model was presented to the world as the only solution to solve the economic problems, eradicate poverty, and guarantee human rights and the rule of law. There is no doubt that globalization did open the markets and lifted hundreds of millions of people out of poverty especially in Brazil, Russia, India, China and South Africa (BRICS) but despite its many gains it was not an equal opportunity provider. Economic inequality increased, in many countries the basic human rights suffered, environmental degradation adversely impacted human health, unemployment soared and communities fragmented in many ways. The unbridled, unregulated free market, cross continental trade, rise of unethical business practices and the emergence of powerful multinationals, unresponsive plutocrats, augmented economic crisis which threatened the well-being of world economies.

The globalization model remained unchallenged for about three decades and unfortunately the intellectual climate of globalization almost eroded our collective sense of exploring alternative models and possibilities. The morbid consequences of exploiting the earth's natural capital wastefully with its concomitants of climate change and distorted and exclusive growth patterns have become reasons enough for the world order to bear the onus of 'rethinking development'. This urge for reassessment resonates in the United Nations 17 Sustainable Development Goals (SDGs) formulated and propagated since January 2016. These SDGs very effectively demonstrate that growth must be both inclusive and environmentally sound so as to reduce poverty and build shared prosperity in terms of health and wealth, today and tomorrow. The only way out of this imminent catastrophe is to find alternative ways of development which are commensurate with holistic sustenance of the life form on earth. The question today is not whether the human race needs to embrace 'Sustainable Development' but 'how' it can do so. The growing disenchantment with the new models of development is generating the gradual realization that everything in the old 'non-developed world', was not so bad. The adoption of 'Yoga' at the international level as a means to 'healthy and good life' is a relevant signpost of this rationale. This could be hailed the beginning of the process to replace the present 'dis-order' by an 'aesthetic order'. If this 'post development' era is to be free of illusions and ideological perversions, hypocrisy and falsehood that pervades it, then it is imperative that the search for signposts and trails leading to the flow of good life be seriously undertaken which should be informed by an entirely new rationale and realistic set of assumptions at this juncture of our 'development' journey. It is crucial that the long list of technologies and technical systems be re-examined from a holistic perspective. Those found incompatible with sustainability and diversity on the planet need to be modified if not wholly abandoned.

The international conference organized in Jaipur by Research Committee-35 (RC- Technology and Development) of International Political Science Association (IPSA) in collaboration with Indian Society of Gandhian Studies (ISGS), Department of Lifelong Learning & Centre for Gandhian Studies, University of Rajasthan, Jaipur provided the requisite opportunity and platform for reflection and discussion regarding the development paths hitherto adopted and those to be pursued for "Sustainable Development".

The conference focused broadly on the following themes:

- Western Models of Economic Growth and Sustainable Development.
- Models of Economic Growth and Sustainable Development in Afro- Asian & Latin American countries (Gandhi, Mao, Mandela, Kwan Lee etc.)
- Corporate Social Responsibility, Globalization and Sustainable Development (issues related to Trusteeship)
- Global Governance and Human Rights: Universal/ Western/ Christian connotations and values
- Development vs Environment Preservation Conundrum: Issues and Challenges.
- Holistic Development: Gender/ Tribal/ Indigenous and other relevant issues

Chapters:

Ten Chapters presented in this compendium address major issues related to main themes of Sustainable Development in the age of globalization. It is interdisciplinary, comparative and provides unique insights into academic and political perspectives of issues.

Introduction of anthology by Rai:

In Chapter I *'Globalization, Climate change, Sustainable Development and Human Security'* Preeti Sharma analyzes the available scientific data on Climate change and its negative impact on human security. – Environment, Quality of life one of the major developments in the 20th century has been the explosion of urban centers around the world in general, and in developing/emergency centers in particular. Migration from rural areas to urban centers has created many social, political and economic problems for the policymakers. How to provide employment, health services, housing, education-the basics of development etc are issues navigated in this paper.

Vishwamaitri Shekhawat and Ashok Kumar Gupta in Chapter II *'Impact of Urban Development on Environment'* discusses the impact of urban sprawl in major cities of India, and the challenges it poses for the general welfare of the population. She suggests several policies to cope with these environmental problems.

In Chapter III, Warby and Pellegrino analyze the role of gender in development in general, and microfinance in particular. They observe that most of these microfinance programs have often targeted women for loans because they empower women (gender equality) as better managers of these loans. The paper *'Do Women in Government Champion Development Friendly Regulations?'* also discusses the role women can play in important political realms facilitating poor low income women to succeed. They observe that 'female legislators and parliamentarians are more likely than their male counterparts to fight for gender issues'.

In chapter IV Satyanarayana Vadlakonda discusses the vital role of multinational corporations in the global economy in *'Corporate Social Responsibility – An Institutional Perspective'* . He presents an 'Institutional Theory' of Corporate Social Responsibility by arguing that the relationship between basic economic conditions and corporate behavior is mediated by several institutional factors. Examples from Indian experiences are offered.

Suman Maurya in her chapter V *'Good Governance as Milestone and Safeguard of Human Rights'* analyses the necessity of good governance in safeguarding human rights in the process of economic, social and political development. Good governance includes transparency and accountability to instill trust among citizens in a democratic context. She cites Indian commitment to guarantee Human Rights in the Directive Principles of State Policy.

Snehil Kacker in chapter VI *'Women as Change Agents of Sustainable Development in South Asia'* analyses the crucial role women play in bringing about change and sustaining it by secularizing younger generations of the family. It has been said that women held the sky but in reality as pointed out by Kacker 'women are more vulnerable to exploitation and torture'. She discusses how societal and political changes ushered by globalisations have contributed to many constructive changes in empowering women. The essay delineates the impact of climate change on women in South Asia.

It also suggests an alternative discourse on the importance of women for Sustainable development.

Chapter VII '*Sustainable Development Goals: Integration of Information and Communication Technologies (ICT) to Deliver Sustainable Development Goals Myth Vs Evidence*' by Sarfaraz Z. H. Khawaja and Akbar Ayesha is a comparative analysis of several examples in Asia, Africa and Latin America with main focus on Pakistan. The data is broadly descriptive and theoretical to provide understanding of the information using content analysis of historical and contemporary review of literature. It aims to analyse the role of smart solutions (SDG) to track students and trackers performance in providing quality education in the overall context of development. The authors prescribe several solutions for effective integration of communication technologies in areas of health, education and general well-being in Pakistan.

Meeta Sharma in chapter VIII briefly discusses the problem of excessive waste created by urbanization, industrialization and the increasing poverty. Issues related to waste management are global but more so in developing countries like India where local bodies lack resources and expertise to properly dispose of millions of tons of waste – a potential hazardous source of diseases. Sharma in her discussion *'Waste Management as a key component of Sustainable Environment'* analyses this environmental degrading problem in India in the global context- efforts of some successful others not so. She concludes that 'Swacchata needs a revolution of ideas, Cleanliness is godliness.'

Chapter IX Radha Kumari in *'UJAMMA: An alternative Model of Development'*, analyses colonial and post-colonial history of Tanzania,its various paths of development after independence and belief that "Capitalist development is incompatible with full national independence for third world countries" hence its own version of socialism as an alternative model for development- the African model of Socialism- the Ujamma model which also did not meet the expectations.

Dhirendra Vajpeyi and Sheila Rai's *'Global Governance and Sustainable Development: Emerging Realities and Perspectives'* in chapter X analyses the nature of Global governance, Globalisation, Democratization and Social capital, genesis of global governance, sustainable development and global governance, and suggests alternative models.

PREFACE

After the collapse of Soviet Union and its Marxist-Leninist economic and political model of development the Washington consensus inspired model of economic and political development, popularly known as Globalization, became the new mantra , the panacea to cure all societal ills, and the only option to raise global humanity from poverty , ill health, and political oppression . This euphoric ideal of globalization which gained popularity in the 1990s, gave rise to an altogether new global order. Promising exciting possibilities and rewarding developments, the process of globalization virtually challenged nations, cultures, religions, markets and people to reorient themselves to the restructuring of the new world. There is no doubt that it lifted hundreds of millions of people out of poverty, however its many gains were not equally or even near equally distributed . Economic inequality, human rights suffered especially in China and several countries in Eastern Europe, environmental degradation, unemployment, unregulated and unbridled dominance by multinationals (mostly Western, Americans) brought about unprecedented economic crisis and social fragmentation. Thus the shrinking world with the implicit hegemony of the 'west over the rest' brought in its wake unprecedented miseries. Amongst the various concerns reverberating across the globe alarms were raised vociferously regarding nation-state's identity and security. Paradoxically these emanated both from the 'winners' and 'losers'. Subsequently we witnessed the reversal of globalization, which manifested in the form of Brexit, and rising tide of protectionism in states which hitherto championed the cause of free trade and liberalization. The "discontent with globalization "forced academics and policymakers to revise, reassess and reevaluate other alternatives.

Realizing the importance of these varied implications of the changing tide when attempts were being made to script new beginnings at the global, regional as well as national level, we as intellectuals considered it imperative to contribute through discussion, debate and research based deliberations towards the augmentation of requisite new horizons of sustainable development.

It is with this purported aim that Research Committee 35-Technology and Development, IPSA, Indian Society of Gandhian Studies, Department of Lifelong Learning and Centre for Gandhian Studies, University of Rajasthan collaborated and successfully organised an international conference in Jaipur on 'Globalization & Sustainable Development: Quest for a New Paradigm' in April 2017. Academics from different parts of the world, belonging to various social science disciplines participated with full vigour in this intellectual churning. Invaluable contributions provided through research papers and insightful deliberations of the participants made the conference very fruitful.

This anthology is an outcome of the international conference. Ten papers have been selected from the large number of research papers submitted on different dimensions of the main subject and other related themes. These papers have been carefully chosen with the objective of providing thematic representation including regional participation. Needless to mention, all the papers selected for this volume have been peer reviewed and revised prior to their inclusion. Due to various constraints, it was not possible to include all the papers submitted in the conference. Nonetheless, we extend our profound gratitude to all the academics and researchers who provided invaluable contributions through their papers and scholarly deliberations and made the conference very successful.

The contributors in this anthology have analysed issues from local, national and global perspectives, attempting to capture the essence of conceptual reflections and empirical realities with reference to topical issues in the context of globalization, exploring alternative models to Western unregulated free market globalization models.

We would like to place on record our heartfelt gratitude to ICSSR, New Delhi for providing financial support to organise the conference due to which we have been able to produce this anthology of scholarly papers. In this context, the University of Rajasthan and the other collaborating institutions also merit deep appreciation for all kinds of support rendered in hosting the conference. It is pertinent to note that the Divisional Commissioner of Jaipur Shri Rajeshwar Singh, who was also serving as Vice Chancellor of the University, was indeed a veritable source of both institutional and intellectual support. We express special thanks to Prof.Vijay Vir Singh, Director of the conference for his intellectual inputs and able guidance in accomplishing the academic aims and objectives of the entire academic exercise.

It is indeed challenging to express indebtedness to all those numerous colleagues, friends and students who rendered unimaginable help and support both in organising the conference as well as in publishing this book. We cannot thank Dr. Preeti Sharma enough, for not just contributing as an author but for also providing unstinted support in all the stages of preparation of this manuscript. Her dedication, patience and good nature tremendously helped in putting together the manuscript from several drafts.

Finally we are very thankful to all the contributors of this volume. Their scholarship has made this book possible. Last but not least we feel very obliged by the Society Publishing, Arcler Education Inc, Burlington, Canada for their painstaking effort in publishing it in this praiseworthy manner.

Sheila Rai
Dhirendra Vajpeyi

Chapter 1

Globalization, Climate Change, Sustainable Development and Human Security

Preeti Sharma

ABSTRACT

Despite the absence of perfectly reliable data about the extent and geographical distribution of climate change on Earth's ecosystem and the quality of human life, there is a broad consensus supported by more than 1250 authors and 2000 scientists from Nobel prize winning Intergovernmental Panel on climate Change (IPCC), as well as 18 American Scientific Associations, that global warming is observably happening and is a growing threat to our planet. Public opinion surveys conducted around the world demonstrate that a large number of general population perceives climate change and global warming as one of the most serious threat to Earth's environment hence to human security.

Following discussion does not present any side of the climate change discourse. It simply aims to provide an objective analysis based on scientific data (climate change models) on the crucial economic, political, and international security issues by addressing THREE questions: (a) Is the climate changing?, and the factors responsible for these changes, (b) Impact of these changes on Earth's environment in general, and human security in particular, (c) What is to be done?

Keywords: *Global warming, climate change, Sustainable Development, Human security.*

During the last fifty years, anxiety and concerns about climate change and its impact on Earth's eco-system and human security have attracted much attention and controversy among scientists, policy makers and other global actors such as non-governmental organizations (NGOs), tribal communities and even the Pope. However, these impact issues have been with us as long as the human civilization itself. In his 800-page Magnus opus book "Global Crisis: War, Climate Change and Catastrophe in the Seventeenth Century" (2014), Geoffrey Parker presents a longitudinal history of wars and climate change, the complex relationship between climate change and global instability. "Climate change has frequently caused or contributed to widespread destruction and dislocation on Earth" (Parker 2014:xx). The adoption of the United Nations Framework Convention on climate change at the Earth Summit in Rio (1992) brought intense discussion on climate change/ global warming at an unprecedented complex Conference attended by 172 nation states, 30000 people with no formal official status belonging to Non-Governmental Organizations (NGOs) of many hues, colors and agendas. The most important contribution of the Conference was that for the first time so many countries "focused attention on the multifaceted and international nature of many environmental and developmental issues" (McDonald 1998:xi). At the Rio Earth Summit "it became very clear that the global environment politics had entered a new stage in which political stakes had intensified and increased" (Koenig 1995:2) but despite this optimism, the Conference also exposed deep chasm and controversies between the emerging economies (the South) and the American led industrialized ones (the North) including Japan on several environmental – developmental issues. (Table 1)

Table 1: Linking Agendas for Sustainability

	The Development Agenda	*The Ecological Agenda*
Northern Agenda	Preservation of wealth, affluent lifestyles and the welfare state	Preventing catastrophe (ozone depletion, green house impact and global warming)
Southern Agenda	Overcoming mass poverty and achieving a higher quality of life	Sustain local environment (stop deforestation, soil erosion and desertification)

Source: Koenig 1995:5

These differences on the impact of climate change on human lives have become entangled into extreme ideological partisan and often times even personal. Passionate supporters and the true believers spew vast amounts of "scientific" data to support the adverse impact of climate change on the quality of human life, and the very survival of the Earth while equally committed opponents have called climate change as a 'hoax', 'a fake science' and a 'crock' ever perpetuated on American public by 'whiners' and 'political hacks' and 'leftists' (Nordhaus 2012:32). Climate change has also been called a "super wicked problem" (Rittel and Weber 1973), "a perfect moral storm" (Gardiner 2011) involving a "conglomerate of complex and at times deeply troubling situations, sometimes referred to as wicked and post-normal" (Rittel and Weber 1973) "with latent and recognized uncontrollable epistemic qualms, catastrophic collective environmental costs and moral, political, financial and methodological inconsistencies of values" (Rittel and Weber 1973).

Keeping in view the above framework the analysis will address the following questions:

- Extent and nature of climate change;
- A brief summary of climate change models;
- Impact of climate changes in general and human security issues, economy, health, agriculture, etc. in particular;
- What is to be done? International response and other policy prescriptions.

(i) **Extent and Nature of Climate Change:** Despite sharp differences between the deniers and supporters of climate change impacts, there is emerging a general awareness about the human induced adverse effects of extreme global warming/climate change on Earth. (Table 2). Extreme climate changes related events in recent years have further contributed to shifts in public opinion. In 2003, summer heat wave in Europe killed more than 70,000 people. In the U.S., hurricane Kartrina destroyed $81 billion worth of property and killed more than 2,000 people and displaced hundreds of thousands of people. Globally, over 106 million people in 2011 were adversely impacted by floods, 40 million by violent storms and 60 million by droughts in Sahel and other parts of Africa. (Parker 2014:xx). Due to change more than 1250 authors and 2000 scientists from the Nobel Prize winning Intergovernmental Panel on Climate Change (IPCC) and 18 American Scientific Associations have observed that global warming is happening and is a growing threat to the world. (UCS Fast Check, November 2011).

A New York Times – CBS opinion poll reported that two-thirds of Americans would back a binding global treaty on climate change (December 1, 2015).

To quote the IPCC, "Our collective security in a fragile and increasingly interdependent world" would be affected by human induced climate changes. Rainfall patterns may be affected contributing to water scarcity, floods, droughts and overall food and health sectors". (The Economist, 2011 May 7:80-87).

Table 2: Climate Change: Opinion Poll

Response	*Countries*	*Percent of Respondents*
Yes	Brazil	86
	India	76
	USA	45
	Russia	33
	Israel	24
	China	15

(The Pew Research Center data cited by Time, November 22, 2015:15)

(ii) **Models are beautiful to look at but...** Scientists have used climate models as "mechanisms to quantitatively measure and estimate the relationship between climate change and climate extremes". (Easterling et al. 2008(a); 2068). Most of these models are "based on well-established physical principles and have demonstrated to reproduce observed features of recent climate and past climate changes." (Randall et al. 2007:59). According to the IPCC "confidence in observed changes in extremes depends on the quality and quantity of data and the availability of studies analyzing these data which vary across regions and for different extremes" (IPCC 2012:6) and there is considerable confidence that these models provide estimate on future climate changes (Randall et al. 2007:600). This perceived confidence in models is mainly based on two following factors:

- Physical bases of the models
- Their ability in representing observed climate, and past climate changes as far back as 6,000 years warm-mid-Holocene or the last glacial maximum of 21,000 years.

However, despite improved and sophisticated technologies to provide more reliable models by General Circulation Models (GCM) serious misgivings about the soundness of these models have been expressed. Easterling et al. (200b:418) point out that there is a shortage of reliable longitudinal data to analyze and measure extremes in climate changes. In most countries "research utilizing high temporal and spatial resolution observations of temperature, precipitation, humidity, winds and atmospheric pressure is relatively new only going back to WWII". Also reconciling ever increasing divergent models has been a daunting task (Reifen and Toumi 2009). Models are beautiful to look at but they are far from perfect and suffer from many technological shortcomings. Critics of GCMs also observe that "the real world data provides little or no support… the global warming of the past few decades was part of the much longer warming trend that began in many places throughout the world a little over three centuries ago (about 1680) with the dramatic beginning of the end" of the Little Ice age well before there was any significant increase in the air CO2 content" (Solomon et al. 2009). Solomon et al. further point out several other "intrinsic" problems and shortcomings with GCMs by setting forth major criteria for modeled climate parameters. (Vajpeyi; 2013:3)

- Observed changes contributed by anthropogenic factors are already happening,
- The Earth's current temperature is no higher (maybe slightly less) than during the peak warmth of the Medieval Warm Period (MWP) when there was more than 100 ppm less CO_2 in the air than there is today;
- There are non-modeled chemical and biological principles that may be equally as important as the physical principles employed in most of the models;
- Projections are available and are broadly robust across models but these models often diverge as greatly in their assumptions and in their specific spatial and temporal findings that they cannot validate each other, nor can such discordant projection be combined to produce meaningful averages… According to studies, "real world data contradict the models… To say that such models are "robust" is wishful thinking" (Woolings 2010:373- 375).
- On the basis of seventeen climate models, analysis by the Fourth Assessment Report of the IPCC no evidence was found "of future prediction skill delivered by past performance based model selection… favoring no particular model or groups of models consistently" (Reifen and Toumi 2009:10).
- Despite these limitations and shortcomings of climate models to perfectly predict climate changes in short and long range there is an emerging consensus that climate "warming is projected to continue through the twenty first century under all models and all emission scenarios and by the end of twenty first century the global temperature will be several degrees Celsius warmer than the present" (Dessler and Parson 2006:81-82).

It must be pointed out here that climate change analysis goes beyond one factor- the rise in temperature. Also Climate is not just temperature but also includes such factors as humidity, precipitation, cloudiness and winds etc." (Dessler and Parson 2006:57) affecting changes in rising sea levels, melting/retreating of glaciers, flow of rivers and water levels of lakes (Vajpeyi 2013:4). Several models supported by new scientific data show that the "destructive effects of climate change have already begun to sweep the planet, with the global economy firmly on track to produce a level of emissions that would lock in a future of rising sea levels, intense droughts and food shortages, more destructive storms and floods, and other catastrophic impacts" (The New York Times, October 16, 2016). According to Dessler and Parson (2006:47) during the twentieth century the global average of surface temperature of the Earth had already increased by 0.4-0.8 degrees Celsius mainly between 1910 and 1945, and then from 1976 to date and 1990s have been the warmest years since climate warming data was kept in the mid-nineteenth century (Figure 1). According to the IPCC (2012) Summary for Policymakers "it is virtually certain that increases in the frequency and magnitude of warm daily temperature extremes and decreases in cold extremes will occur in the twenty first century at the global scale and can result in unprecedented extreme weather and climate events" (Figure 2). According to The Economist (May 28, 2016:70) "between 1998 and 2013 Earth's surface temperature rose at a rate of 0.04C a decade, rather than the 0.38C increase of the 1990s". The "refusenicks" and deniers, however, assert that most of the climate changes' dire predictions are simply scare tactics, and a modest warming is likely to be beneficial. "We do not see any significant warming of the atmosphere…the oceans covering 71% of the Earth's surface shows no appreciable warming either." (Singer 2012:85)

The complexity and uncertainty about the extent and nature of climate change is due to not only the interpretation and paucity of longitudinal data but is also driven by political/ideological factors. "A debate….like almost all environmental issues…a debate over culture, world view and ideology" (Hoffman 2012:30-37). Absence of consensus on climate change "strengthens its opponents/deniers and drives green-minded politicians mad." (The Economist: Special Report on Climate Change, Hot and Brother, November 28, 2015:5)

Various climate models point out that changes are already taking place and there is sufficient evidence for anthropogenic contributions to these changes (Solomon et al. 2009) as evidenced by following developments in nature.

(a) **Oceans:** Oceans cover 71 percent of Earth's surface and are good indicators of global warming. Dessler and Parson (2006:53; 65) point out that during the twentieth century global average sea levels rose by 1.5mm per year of 15cm over the last century, and since 1970's the sea has retreated by around 12 percent each decade. The rise in the sea levels is already having adverse impact on low-lying areas of

Maldives, Bangladesh and United States (The New York Times 2016: November 22; March 15 and 11, 2012). It could affect an estimated 145 million people around the world (*Hertsgaard 2011:33*)

(b) Glaciers: Glaciers cover about 10% of the Earth's total area and are one of the major providers of water. They are mostly in Greenland, Himalayas, and Antarctica. Any substantial rise in temperature will melt (retreat) these glaciers pumping a large amount of water in rivers and also raise the sea levels. Rising water in rivers will flood precious agricultural land and human habitats but also in long range deprive much needed water for consumption. It has been reported that between 1860 and 1900 thirty-six glaciers were observed. Thirty-five of them retreated due to rising temperatures. One hundred and forty-four more glaciers were monitored between 1900 and 1980. A total of one hundred and forty two retreated due to warming temperatures. "The Arctic is warming roughly twice as fast as the rest of the planet. Since 1950's, the lower atmosphere has warmed by a global average of 0.7 degrees Celsius. Greenland's air has warmed by 1.5 degrees. The main reason appears to be catalytic warming effect triggered by global warming." (Astill 2012:5) In Greenland and Antarctic "massive ice-sheets " have started to collapse creating "massive ice-sheet instability "which were not reported by pre-Paris conference on Climate Change sea level rise models. According to more recent scientific data "the collapse of Greenland's ice-sheets will have a greater effect in the southern hemisphere, while collapse in Antarctica will have a greater effect in the north, the phenomenon known as "regional effect fingerprinting " (Crist 2018:12). "There is no longer any doubt that the rise in the global sea levels will reshape human civilization". (Goodell 2018)

(c) Precipitation Patterns: Most of the scientists generally agree that Earth's climate is changing due to "natural variability" and increased concentrations of greenhouse gases and aerosols" (National Aeronautic and Space Administration) and would profoundly impact atmospheric water vapor concentrations, clouds and precipitation patterns, an increase in the amount of moisture circulating throughout the troposphere (lower atmosphere) leading to an increase in the intensity of precipitation primarily over land and thus more rainfall instead of snowfall in some regions of the world (NASA). Van Aalst (2006:9) and the IPCC (2012) point out some of the risks.

- Increased floods, landslides and mud slide damages;
- Increased soil erosion;
- Increased floods run off;

- Increased pressure on governments and private flood insurance systems and disaster relief.

These perceived risks can damage crops, contaminate water supplies, and spread infectious diseases such as cholera, and damage infrastructure and transportation.

According to the IPCC (2012:6) "there have been statistically significant trends in the number of heavy precipitation events in some regions of the world." However, this increase has been uneven in different parts of the world (IPCC 2012: Flannery 2006:124). The IPCC (2007:7) reports that "north and south America, northern and central Asia and northern Europe" have experienced an increase in the amount of precipitation between 1900- 2005. Global precipitation has increased by 7 percent over the last fifty years (Huber and Gulledge 2011). The IPCC (2012:7) has asserted "medium confidence" (a five in ten chances) that anthropogenic activities have been a factor in the increase of intense precipitation events." The IPCC (2012:11) predicts that it was likely that the frequency of heavy precipitation of the proportion of total rainfall from heavy falls will increase in the twenty-first century over many areas of the globe". It will increase predominantly in "high altitudes and tropical regions and in winter in the northern mid-latitudes" (IPCC 2012:11). "Based on a range of emission scenarios, a one in-twenty-year annual maximum daily precipitation amount is likely to become a one-in-five to one-in-fifteen year event by the end of the twenty-first century in many regions and in most regions the high emission scenario leads to a stronger projected increase". (IPCC 2012:11).

(d) **Monsoons:** According to IPCC (2007) "it is likely that warming associated with increasing greenhouse gas concentrations will cause an increase in Asian summer monsoon precipitation variability." According to Houghton (2001:16), "climate change can contribute to changes in strength and lifespan of monsoons by increasing or decreasing these features." For example, the Mexican monsoon season is already beginning about ten days later than usual, and although there are fewer rain events in Southwestern North America, the event that do occur are more intense. (White et al. 2013:32, Karl 2008:4). It has also been observed that "South Asian summer monsoon" could come later in the season and be weakened due to rising temperatures… "and as a result" agricultural production, water availability and hydroelectric power generation could be substantially affected by delayed monsoon onset and reduced surface runoff and contribute to increased rainfall and flooding in Bangladesh posing great risks to local communities (USA Today 2009: February 7). The IPCC (2012:13-14) however, states that "confidence is low in projection of changes in monsoons (rainfall) because there is little consensus in climate models regarding the sign of future change in the monsoons".

(e) **Floods:** It seems logical that melting glaciers and increased rainfalls in some regions of the world could contribute to floods and impact human lives. However, the IPCC (2012:6) reports show "limited to medium evidence available to assess climate driven observed changes in the magnitude and frequency of floods at regional scales because the available instrumental records of floods at gauge stations are limited in space and time and because of confounding effects of changes in land use and engineering" and it is difficult to determine the difference between "freak events" and "real changes in the probability of a particular event" (Van Aalst 2006:11). These contradictory and conflicting data confuse the issue and create more inconsistencies on the topic.

(f) **Droughts:** Mother nature has not been consistent. The available data show increased precipitation intensity in several regions of the world while serious decrease in others creating serious drought conditions. It has been observed that "drought is one of the most costly types of extreme events and can affect large areas for a long period of time" (Karl 2008:5). IPCC reports consistently point out the potential impact of droughts on human security such as farmers and food shortages. Droughts have potential to contribute to a reduction in crop production, damage to building foundations caused by ground shrinkage, a reduction in water supply both in "quantity and quality" and in increased probability of forest fires (Van Aalst 2006:9; Vajpeyi 2001:1-21).

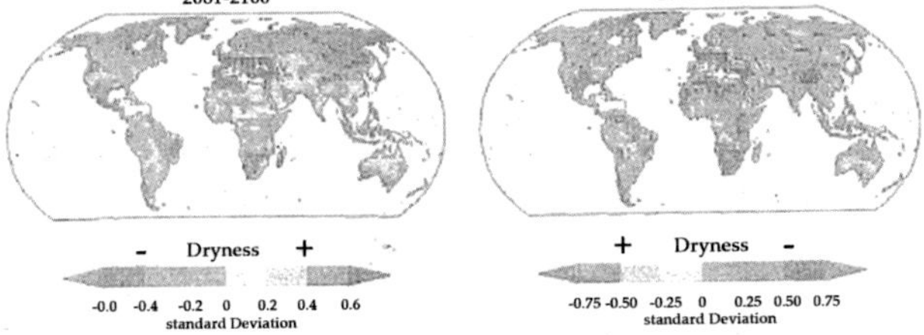

Figure 1: Climate Change, Precipitation and Dry Days

Source : Special Report, IPCC 2012 : 13

"Since the 1970s, droughts have been reported in large areas in tropics and sub-tropics in Mediterranean, Southern Africa and parts of Southern Asia (White et al. 2013:34). Overall, drought conditions in Africa in general and the Sahel region in particular have turned the region as "dust filled wasteland." It has been argued that climate changes have also been affected by reckless human activities (Giannini et. al. 2003; Flannery 2006:124; Zeng 2003) and the rising temperature of the Indian Ocean, collection of greenhouse gases was one of the major factors for the "dramatic climate shift." (Figure 1)

(g) Tropical cyclones: Tropical cyclones are usually created over oceans with temperatures of at least 25.5C. They can last for at least two weeks but quickly lose momentum as they head towards inland creating immense property and human lives damage (White et. al. 2013:35) such as:

- Damage the crops and trees
- Power outages, disruption of public water supply
- Deaths, injuries, and water borne diseases
- Damages to buildings and other infra- structure.

Tropical cyclones are difficult to predict despite sophistical satellite technologies due to their infrequent occurrences and 'high inter-annual and multi-decadal variability in tropical cyclone occurrence' attributed to the 'El Nino/ Southern Oscillation (ENSO)', the stratospheric quasi- biennial oscillation and multi- decadal oscillation in the North Atlantic (Van Aalst 2006:13 in White 2013:35). It has also been observed that "blaming climate change for particular storms remain tenuous."(The Economist, May 28/2016:7)

(h) The Ozone Layer: The Ozone is a gas in the upper atmosphere. It protects the earth from Sun's radiation. The average Ozone layer is about 3 millimeters thick or 300 Dobson units while the Ozone Hole (the depleted area) averages around one millimeter thick or 100 Dobson units. The Ozone layer is depleted by the over use of Chlorofluoro Carbon (CFCs) used in aerosol cans. Scientists and Climatologists found serious depletion of the Ozone in 1980s over Antarctica causing serious concern about the health and well- being of humans and animals alike. The global community adopted the Montreal Protocol in 1987 to reduce and eliminate the substances released in the atmosphere Ozone Depleting Substance (ODS) to protect the Ozone layer. The Protocol required that the US must phase out the production and use of ODS by 2030. Developing countries were to phase out these substances by 2040. The deadline for the European Union (EU) was set at the year 2020. The European Union met its obligations ten years ahead of its deadline (2010) and also regulated chemicals 'not considered as consumption under the Montreal Protocol' (Climate Action, March 11/2011). In October 2016, in a major breakthrough more than 170 countries agreed in a legally binding agreement in Kigali to cut the world wide use of planet warming chemicals used in air- conditioners and refrigerators. President Obama called the Kigali Accord as "an ambitious and far reaching solution to this looming crisis". (The New York Times, October 16/2016)

(i) Carbon Di-Oxide and Climate Change: The Industrial Revolution in Europe (mid- eighteenth century) drastically impacted economic and political spheres. The Revolution introduced new ways in manufacturing and labor practices. The cheap coal powered these industries which emitted vast amounts of CO_2 in the atmosphere. By 2005 the presence of CO_2 had reached alarming proportions. "Global atmospheric concentrations of CO_2 were 35 percent higher than they were before the Industrial Revolution" (The EPA, April 14/2011). Excessive presence of these CO_2s affects climate and human health by polluting air, and water supply.

Most of the world's greenhouse gas emissions come from a relatively small number of countries. China, the United States, and the nations that make up the European Union are the three largest emitters on an absolute basis. Per capita greenhouse gas emissions are highest in the United States and Russia. (Figure 2)

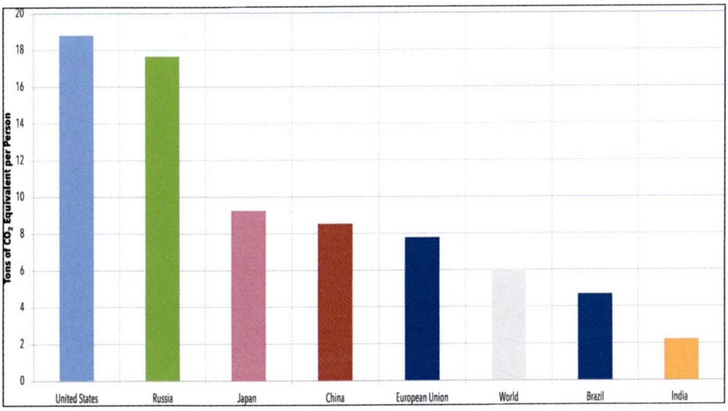

Figure 2: Per Capita Greenhouse Gas Emissions, 2017

Source: "Per Capita GHG Emissions" Centre for Climate and Energy Solutions (https://www.c2es.org/content/international-emissions/)

These Green House gases are absorbed by so called 'sinks'. The EPA points out following ways CO2s are emitted and removed. (The EPA, April 14/2011; White et. al. 2013:40-41)

- Plant and animal respiration-'Oxygen and nutrients are converted into CO_2s... Plant photosynthesis removes these CO_2s from Earth's atmosphere'
- Oceans are vast 'sinks' to absorb these Greenhouse gases.
- Volcanic eruptions emit the Carbon.

Recent data from EPA (April 14/2011) points out that these natural 'sinks' are overworked and therefore unable to absorb these gases released by human, and animal activities. Several suggestions especially geo-engineering, have been made to reduce the adverse effect of CO2s on Earth's atmosphere. The biggest source of CO2s has been the use of fossil fuels. Increased use of fossil fuels have "over the past century increased the global sea level by close to 17 centimeters, but within the last decade this rate has nearly doubled" (Figure 4) increasing sea levels and threatening low lying areas around the world. "With temperatures rising at a rapid rate the ice- sheets have been melting". Between 2002 and 2006 Greenland's ice- sheets have been depleted by 36 to 60 Cubic miles per year and measurements show that "Arctic ice has rapidly declined over the past several decades" (White et al. 2013:43).

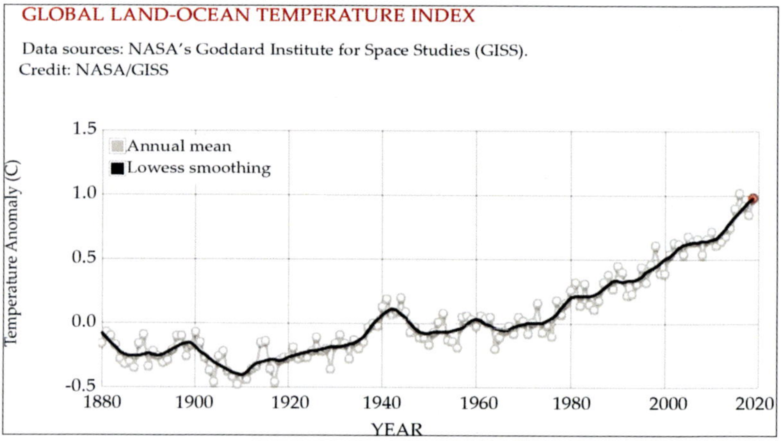

Figure 3: Global Land-Ocean Temperature Index 2019

The New York Times (November 29/2016) reported that "climate change pushes towns in Alaska to wrenching choice". With its proximity to the Arctic, Alaska is warming about twice as fast as the rest of the United States and the states are heading for the warmest year on record. The government has identified at least thirty one Alaskan towns and cities at imminent risk of destruction… Climate change experts predict that these will be uninhabitable by 2050, their residents joining a flow of climate refugees around the globe, in Bolivia, China, Niger and other countries". It must be noted that "between 1998 and 2012 humans have pumped unprecedented quantities of greenhouse gases into the atmosphere without raising average global temperature because much energy went into melting ice, absorbed by oceans." (The Economist, November 28/2016:5; Parker 2014:687)

Data source: NASA's Goddard Institute for Space Studies (GISS), "Global climate Change, Vital Signs of Planet" NASA (https://climate.nasa.gov/vital-signs/global-temperature/)

Keeping in view the above discussion, despite differences of opinions, on the causes and extent of impact of human activities on climate change, there is a strong set of data supporting the climate change from China to Peru (Figure 4) and it poses severe threat to the overall well-being of the Earth, human security and sustainable development (Figure 5)

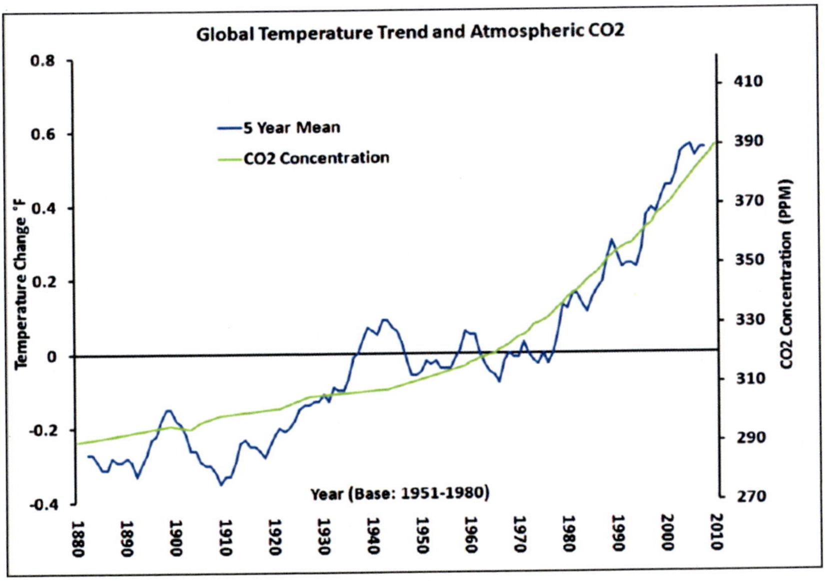

Figure 4: Atmospheric Carbon Dioxide and Global Surface Temperature Trends

Source: Center for Climate and Energy Solutions (http://www.c2es.org/facts-figures-trends/CO₂-temp)

CLIMATE CHANGE HUMAN SECURITY AND SUSTAINABLE DEVELOPMENT: A PARADIGM SHIFT:

Human Security

Due to the immense complexity of human society solutions to its problems require a multi- disciplinary and multi- dimensional approach. Several scholars maintain that human security concept is too complicated to be a 'practical guide' for research, and to be properly operationalized. It is "extraordinarily expansive and vague, encompassing everything from physical security to psychological well-being." (Paris 2001: 88; Dwivedi 2008; 60-75). Undoubtedly, ideological and practical justifications have made it quite difficult to come up with a sound definition of the concept; however there has emerged a broad based consensus spearheaded by concerned international agencies such as the United Nations, sovereign states, NGOs and citizens. Defining human security has been like asking ten blind persons to describe an elephant. Each person based on his/her experience had a different notion of the animal. According to King and Murray (2001: 589-598) human security is characterized by four main characteristics:

- It is universal,
- Its components are interdependent,
- Ensured through prevention,
- People oriented

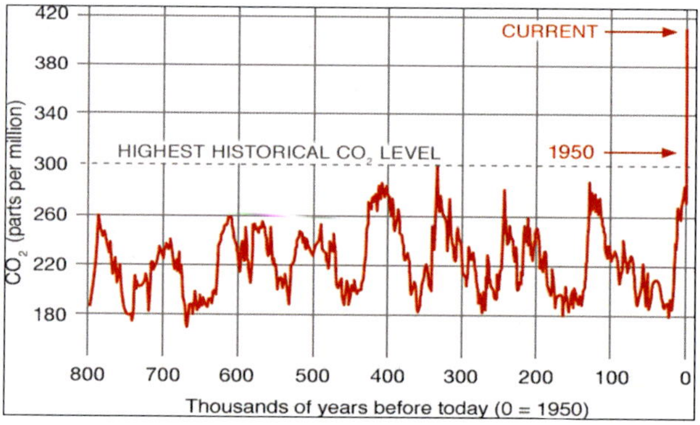

Figure 5: Atmospheric CO_2 Levels

Source "Global climate Change, Vital Signs of Planet" NASA (https://climate.nasa.gov/vital- signs/carbon-dioxide/)

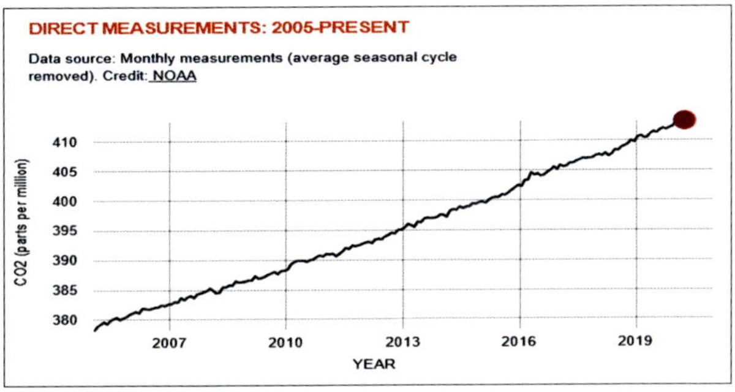

Figure 6: Atmospheric CO2 Levels

Source "Global climate Change, Vital Signs of Planet" NASA (https://climate.nasa.gov/vital- signs/carbon-dioxide/)

The UN Commission on Human Security in its 2003 report 'Human Security Now' observed that the international community urgently needed a new paradigm of security. It presented following six elements as part of human security agenda.

- Conflict Prevention
- Disease eradication
- Poverty Alleviation
- Sustainable Economic Development
- Food Security and
- Promotion of Human Rights

Subsequent United Nations sponsored conferences including its Millennium Summit supported the main goals of the earlier United Nations reports on human security adding freedom of speech, association, religion and movements.

Amartya Sen felt that health, literacy, gender equality and respect for fundamental rights must be essential part of human security. Myers (1993:31) observed that "in essence… human security applies most at the level of the citizen. It amounts to human well-being not only protection from harm and injury but access to water, food, shelter, health, employment and other basic requisites that are due to every person on earth." Dwivedi (2008: 65) observed that human well- being should also include 'social regrettables' such as pollution, crime and divorce, social cohesion (social capital), better health, and the quality of the environment. Deaton (2013:24) maintained that well being refers to all of the things that are good for a person, which make for a good life … includes material well-being such as income and wealth; physical and psychological well-being represented by health and happiness, and education and the ability to participate in civil society through democracy and the rule of law." An earlier UNDP Publication (1994) on Human Development observed that "at the global level human security no longer means careful constructed safeguards against the threat of nuclear holocaust ... Instead it means responding to the threat of global poverty, travelling across international borders ... Climate change, illegal migration, and terrorism ... and the threat of global poverty ... is real and persistent." Matthews' definition mainly focused on global issues such as stratospheric ozone depletion, greenhouse warming and deforestation (1989: 162-177), Brown in his definition of human security includes health, climate, governance, energy, human rights, and depletion of ozone layer (2006: XXIII-XXVII), Nef. (2008: 141-169) presents five categories to define human security (insecurities):

- Ecological security
- Economic security
- Societal security
- Political security
- Cultural security

The Stockholm Conference on Human Development, sponsored by the United Nations Development Program in 1972, broke new grounds by emphasizing the quality of human life and its relationship with poverty and ecology. In 1992 the Earth Summit held in Rio de Janeiro, Brazil was one of the earliest international conferences which pushed the notion that economic development mainly measured in per capita income growth without ecological and human security concerns does not contribute to the overall economic and social development, and sustainable growth. The Earth Summit's crown jewel product was a 400 page document the Rio Declaration on Environment and Development popularly known as Agenda 21. It's Preamble proclaimed for the need of a concentrated effort to overcome global hunger, poverty, ill-health, illiteracy, andthe declining environmental health of the Earth in four major themes in forty separate sections, and onc hundred separate action programs interconnectedness between environment and economic development. It also emphasized on the necessity to forge a global partnership between industrialized/developed and emerging economies to face environmental and human security issues. The Agenda 21 encompasses several crucial aspects related to basic human needs such as food, shelter, healthcare, education, culture and human rights. Special attention was given to children, women and indigenous people. Ethical issues, the role of NGOs, military, democracy, and the capacity building of national and international institutions in the context of human security and overall social and economic development were important items (Vajpeyi 2006:92-132). Above discussion of major point of views, reflects the categories presented by UNDP's Human Development Report (1994) as summarized by Paris 2001:90, and Henk 2005:91).

- Economic Security (Freedom from Poverty)
- Health Security (Protection and Prevention from diseases)
- Environmental Security (Climate change, water resources, pollution, exploitation of natural resources)
- Food security (access to basic foods)
- Personal security (Protection from violence, torture, domestic violence, drug addiction, safe habitat)
- Community Security (Protection of traditional cultures, values and ethnic-religious violence)
- Political Security (Civil and Political rights, rule of law, freedom from repression, human rights).

Several other human security concerns such as war crimes, genocide, ethnic cleansing and mass- atrocities have been added (Welsh 2009:4).

RESPONSIBILITY TO IMPLEMENT AND PROTECT THE FUNDAMENTALS OF HUMAN SECURITY: HUMAN SECURITY WITHOUT BORDERS:

Most political thinkers from Plato, Hobbes, Locke, to Kautilya have justified the legitimate creation of the state as the sole agent with coercive power to provide basic security to its citizens.... be an utopian or a totalitarian system (life, liberty and the pursuit of happiness) by impressing them that these programs are there to help them in meeting their responsibilities and not to dilute their sovereignty (Welsh 2009:4).

Globalization and Human Security

Following is a brief discussion of two main components of human security.

- **Health:** According to Jian Li "conceptualizations what it means to be healthy are ultimately shaped by our economic standings, geopolitical locations, able bodiness, age and sex" (2013). The World Health Organization maintains that health is "a state of complete physical, mental and social well-being and not merely the absence of disease or infirmity." Angus Deaton in The Great Escape, Health, wealth and the Origins of Inequality (2013:26,28) observes that "health is not just a matter of being alive, and living a long time but living in good health … High life expectancy, good health, and absence of poverty, democracy and the rule of law are among the features we would include were we to design an design an ideal country." Deaton also points out that "there is a positive correlation between health and income. For much of the world, proportional increases in income are associated with the same increase in life satisfaction" (2013:33). Good health also has security implications, United Nation Security Council Resolution 1308 (2000) addressed the impact of HIV/AIDS in Africa, and its impact on peacekeeping missions, and economic productivity. The impact of globalization on health had both positive and negative sides. "The past half-century has seen an unparalleled integration of the world- a process often referred to as globalization. "Globalization has affected health in many ways directly through the spread of disease, information, and treatment, and indirectly through economic forces, particularly increased trade and higher economic growth. However there are still persistent inequalities and millions of people around the world have no access to drugs." (Deaton 2013: 149-150) In cities like Delhi, Johannesburg, Mexico City, and Sao Paulo, first-world state-of-the-art medical facilities treat the wealthy and powerful, sometimes within sights of people whose health environment is not much better than that of seventeenth-century Europe (Deaton 2013:152). Unfortunately despite international efforts by private foundations such as Gates Foundation, United Nations and NGOs a larger number of populations are denied basic medical facilities for cultural/religious reasons. Nigeria and Pakistan, for quite a while, did not allow its population to have Polio injections for religious reasons (non-health reasons) and "these 'health inequalities' are one of the great injustices of the world today" (Deaton 2013:7).

- **Free Speech:** Freedom of speech and expression are two most important sustaining pillars of democracy. Globalization through open markets, new advances in information technology and social media provided an immense boost to free speech around the world and has "led to a vast increase in the volume of material published" (The Economist, June 4th 2016:59) and ideas exchanged on all sorts of political and cultural topics, opening up new opportunities for innovation and creative dialogues. But it has also faced unprecedented censorship not only in totalitarian countries like China but even in India a functioning democracy. Other governments such as Pakistan, Bangladesh, and Saudi Arabia censor free speech for cultural, religious and national interests. A recently noted threat to free speech is not only from governments but from non-state actors. "Globally the willingness of some Muslims to murder people they think have insulted the Prophet has chilled discussion of one of the world's great religions- even in places where Muslims are a minority such as Europe. Radical Islamists are attempting to enforce a global speech code, in which frank discussion of their beliefs is punishable by death" (The Economist, June 2016: 55-56). The article 'Free speech- the muzzle grows tighter' (The Economist, June 2016: 55-58) opined "that free speech is in retreat. Granted, technology has given millions a megaphone, and speaking out is easier than it was during the cold war, when most people lived under authoritarian states", however, the share of the world's population living in countries with a free press fell from 38% in 2005 to 31% in 2015 … Reporters without Borders' global index of press freedom has declined by 145 since 2013. It is "Shut up or I'll kill you" syndrome.

- **Environmental and Food Security:** Despite some deniers there is a broad census among environmental scientists and policymakers that our collective security is dependent on a healthy environment. The impact of globalization on environment and food security presents a glass half full and half-empty scenario. Today's globalization has lifted hundreds of millions of people from abject poverty to prosperity but it has also produced income and wealth inequalities. Transfer of outdated polluting technologies from rich countries to poorer and less developed economies such as China. China's prosperity came at a high price. Today China is the largest emitter of carbon dioxide (CO_2). Water and air pollution choke the environment. Similar trends have been noticed in India, Brazil, Bangladesh and several other countries. As a result higher temperatures, more frequent heavier rains, stronger hurricanes, increased drought, delayed monsoon, and melting glaciers will affect agriculture and food production threatening food security mainly in the poor countries and poorer people; (Jian-Li 2013:111-138; Assad et al.,2013: 169-200; Oberoi and Singh 2013: 201-226; Cheng Fang Ting 2013: 227-248; Grigoryev 2013: 249-280; Ermina and Barygin 2013: 307-324). Naomi Klein laments (2016-13) that "theories of human hierarchy- that we must take care of our own first- will be marshaled to rationalize these monstrous decisions … Although climate change will ultimately be an essential threat to all of humanity, in the short term we know that it does discriminate, hitting the poor first and worst, whether they are abandoned on the rooftops of New Orleans during Hurricane Katrina or whether they are among the 36 million

> who according to the UN are facing hunger due to drought in Southern and East Africa."

To conclude it is not an easy task to arrive at a consensual framework of human security in the context of globalization. It is more difficult to implement due to many political, economic and cultural factors. In 2016 despite valiant efforts by many governments, UN and non-government actors thousands of Yazidi Women in the Middle East, are being enslaved, in Nigeria young girls are abducted, raped and killed, in Afghanistan they are denied opportunities to get education, and in Bangladesh secular voices are muffled and people killed. A society with even semi-perfect human security rights is still a dream and a fantasy. Religious, ethnic and national identities still drown the integrative values hoped by globalization.

Climate Change and Sustainable Development: Sustainable Development is a multi- dimensional and complex concept. Perceptions and operationalization of sustainable development are influenced by cultural, political, economic and geographical differences around the globe, however, one constant desire/ dream of humans have been to live in a relative secure environment but also in a society where basic needs of daily life are available- enough food, clean water and air, housing etc.- in the ultimate pursuit of life, liberty and happiness. Many models from Marxism to Capitalism, modernization, globalization and changes in global value systems have immensely impacted the articulation and perception of human security and sustainable development. The post WWII kindled hope and optimism to live in peace and prosperity, however it did not take much time to start a money and resource depleting rivalry between the US and Soviet Russia. Narrow national interests explained in competing ideologies squandered precious resources in militarization ignoring vast needs in most of the underdeveloped world mainly in Africa, Asia and Latin America. Quite often development projects sponsored and financially supported by multilateral aid agencies such as the World Bank and other regional Banks to improve economic and natural resources proved to be the "direct or indirect causes" (Caldwell 1984:172, Vajpeyi 1995:23-45). Smog covered cities, acid rain, and thousands of tons of hazardous industrial waste in industrialized countries became painful reminders of potential ecological and environmental apocalypse, and sources of pollution and environmental degradation around the world. It took exceptional determined conscience raising efforts by national and human resources (Caldwell 1984:173). By late 1980s major international actors- US Agency for International Development (USAID), International Monetary Fund (IMF), the World Bank, the Organization for Economic Cooperation (OEC), the United Nations and national governments both in the industrialized West including Japan and the Third World started to reevaluate their past policies and approaches to sustainable development and its relationship with environmental issues. This paradigm shift was partly due to domestic and international pressures exerted by public opinion, and realization that economic development mainly measured in per capita income growth without other crucial factors essential to improve the quality of human life both in short and long range, has resulted in unbalanced distribution and control of economic wealth. In 1987 with the publication of the World Commission on Environment and Development (WCED) Report. **'Our Common Future'** these crucial issues of development and environment were thrust at the global stage. "It was one of those paradigm shifts evolving underneath the surface, visible to a few researchers, insiders, and activists, when it finally reached the mainstream political agenda" (Koenig 1995:2). **'Our Common Future** (Brundtlandt Report) 'de-ideologized' international relations and "finally opened the doors for the kind of political, economic and technological cooperation necessary to pursue an economically sound development path" (Koenig 1995:2). This paradigm shift took place from a model, called 'development' to one called 'sustainability'. As mentioned above that to define 'sustainable', 'sustainability' has not been easy due to the multi-dimensional nature of the concept, however, the Report provided definition of sustainability as "development that meets the needs of the present without compromising the ability of future generations to meet their own needs" (MacNeill et al. 1991:29). According to Koenig (1995:3) "the concept of sustainability is a response to the recognition that current policies are irrevocably undermining our resource base and present biodiversity… the new terminology aims to integrate economic and development strategies". The approach is also termed as 'GAIA model' that "we are part of Quasi- living entity that has the capacity for global homeostasis" (Lovelock 1987:19). In **'The Beyond the limits'** Meadows et al. (1992) it was pointed out that while the economic system is trying to achieve growth, the earth system is finite and non-growing.

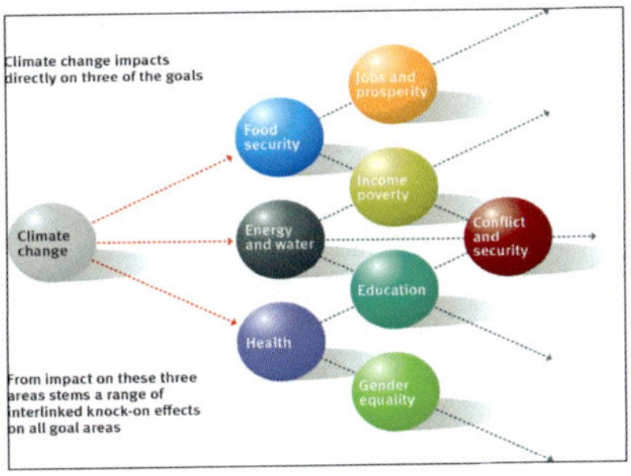

Figure 7: Agreement on Climate Change and Sustainable Development

Source : Rebecca Willis (2014) Paris 2015: Getting a global agreement on climate change.

The ‘carrying capacity’ of the Earth’s system is limited. The Brundtland Commission was mandated to prepare a global agenda for change. In 1989 the United Nations General Assembly by a Special Resolution called on the leaders of the world to meet in three years to explore new ways to sustainable development and environment. Heeding this call the United Nations Conference on the Environment and Development (UNCED), known as the Earth Summit was convened on June 13/1992 in Rio de Janeiro, Brazil. It attracted the “largest and most complex conference ever held” (Elmer- Dewitt 1992:42). The most important contribution of the Conference was that 178 nations, despite several serious disagreements between the rich industrialized countries and poor countries of the Third world, adopted twenty seven non- binding provisions related to sustainable economic development and environment. Oberoi (2017) in her analysis on ‘Sustainable Development and Globalization’ believes that “the agreements on climate change and SDGs should be seen as complementary, with opportunities for mutual benefit in areas such as low carbon development, climate adaptation and resilience and new flows of finance” (Figures 7 and 8)

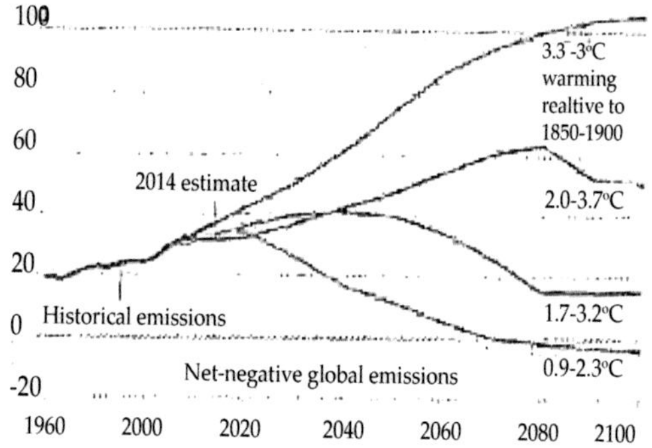

Figure 8: The Course of Global Emissions Four Scenarios

Source : Guardian, 2013

The Earth Summit’s crown jewel was a 400 page long document popularly known as **Agenda 21**. Its Preamble proclaimed the need for a concentrated effort to overcome global hunger, poverty, ill-health, illiteracy and the declining environment of the Earth with an emphasis on inter-connectedness between environment and economic development. The **Agenda 21** emphasized social development by co-opting women, farmers, native populations in conserving and managing resources for sustainable development, rely on efficient use of energy resources by using more renewable non- polluting energy, ozone depletion and global warming. “The Climate Change denies to go right on denying humanity caused climate change in the face of rising waters, fires, floods, droughts and unprecedented storms" (Miller 2016:10). There has also been tremendous response around the world to meet these climate change related challenges.

Following is a brief summary of climate change:

"Simultaneous impact of climate change and globalization" will expose "certain regions, sectors, ecosystems, and social groups" (O'Brian and Robin M. Lei-Chenko 2000:221). Social and economic impacts will be uneven on rich and poor countries (Mendelshon et al. 2000: 221-232; Grubb 1995: 463-496; Parry and Carter 1998). However recent climate change related events show that they do not discriminate between rich and poor, Hurricane Katrina in the US, Tsunami in Asia, nuclear meltdown in Japan due to Tsunami impacted all sectors of society (The New York Times 2012: March 14: A1 and 3)

- Increase in the number of people suffering from death, disease and injury from heat waves, floods, storms, fires and droughts.. and infectious disease vectors.. Diarrheoa diseases. Cardio respiratory morbidity and mortality associated with ground level zones, risk of Dengue (IPCC 2007: II: 393)
- Scholarly studies support above observations in several country studies on the impact of Climate Change on all spheres of human life (Li Jian 2013:111-132); Assad et al. 2013:169-199; Oberoi and Singh 2013: 201-226; Fang- Ting 2013: 227-247; Grigoryev et al. 2013: 249-280; Williams 2013: 281-305; Ermina et al. 2013: 307-321; Mishra et al. 2013: 325-335; Janeh 2013: 51-82; White et al. 2013: 27-50;Johnson 2013: 83-110) in Brazil, India, China, Japan, Turkey, Bangladesh, Nepal, Russia and others. (Vajpeyi 2013)

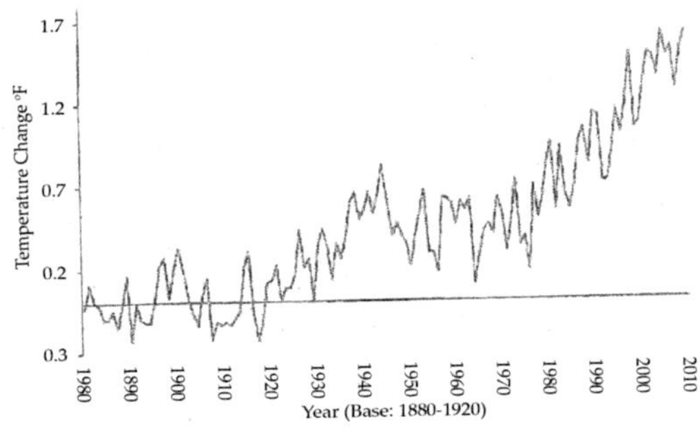

Figure 9: Global Surface Temperature Trends

Source : Center for Climate and Energy-Solutions (http://www.c2es.org/facts-figures/trends/surface-temp).

- Developing countries will suffer more than the developed ones. Most of developing countries are in tropical regions, more dependent on agriculture and other natural resources, have fewer financial resources, weak social and economic infrastructure to successfully cope with drastic changes (Mendelsohn et al. 2006: 163-171: Grubb 1995:467) Gaby (2011:1) further presents several other developments due to climate change.

- Higher temperatures and more hot days in the 21st century (Figure 9)
- Increased frequency of rainfalls especially in winter. (The Economist 2016: November 12:72)
- Stronger Hurricanes (Parker 2014: XX)
- Increased droughts
- High sea levels eroding more coastal areas.
- These changes will adversely impact agriculture, water supplies, health, and even tourism, and global economics.

These observations have been challenged by climate change deniers including Americas president elect Trump. "Globally, the debate on sustainable development and climate change has been a fertile ground for mutual suspicion between the North and the South.." (Oberoi 2017)

International Response

Despite differences of opinions about the nature and extent of climate changes on our lives vigorous debates have kept the issue at the forefront. Hundreds of international and national conferences sponsored by aid agencies, the United Nations, green movements led by NGOs, scholarly research, and active participation of policy- makers on both sides- opposition and support of the discourse are clear indications that the issue is crucial for the very survival of human civilization. Traditional theories of economic development have been challenged and a paradigm shift on environment and sustainable development is becoming part of the mainstream consensus from Russia, China, and Brazil to India, South Africa and other countries. Following is a brief discussion of only major landmark attempts in the environmental context.

The United Nations Conference on Human Development held in Stockholm Sweden in 1972 was the first major attempt to accept the nation that there is a close relationship between poverty, poor health, human rights, human security and a balanced economic development. The Conference adopted a Declaration of Principles emphasizing the need for global cooperation and action to tackle the problem. The progress to translate these principles into a action plan was slow due to lack of financial resources, serious differences between poor and rich countries- mainly who is responsible for environmental degradation, and who should pay to clean it up (Polluter should pay), transfer of green technology. Several disturbing ecologically destructive events in different regions of the world- Bhopal tragedy in India, the Love canal toxic waste in New York, nuclear waste dumping by Italy in Nigeria jolted the public conscience on irresponsible behavior of multi- national corporations such as Exon, Dow Chemical etc.- and sovereign states in the West. It was felt that pious resolutions must be followed by concrete actions, and guilty parties must be publicly exposed and punished. As discussed earlier that the United Nations General Assembly by a special mandate appointed a Commission- the Brundtland- to come up with practical ideas and a framework to address this complicated issue.

The publication of Brundtland Commission report **'Our Common Future'** in 1987 not only introduced the concept of Sustainable Development but clearly observed a close and crucial relationship between Sustainable Development based on a new paradigm. It was at the Earth Summit in Rio that two major treaties were approved:

- The UN Framework Convention on Climate Change which addressed global warming issues,
- The Convention on Biological Diversity to protect Ozone depletion.

The 190 signatories at Rio agreed to evaluate their greenhouse gas emissions and ways to reduce them. Between Rio (1992) and Kyoto (2005) the growth of the climate change regime has been divided by some scholars into periods (Oberoi 2017; Bodansky et al. 2015)

- Foundational phase: Development of scientific concern on the climate change
- Agenda setting phase (1995- 1988): transformation of the issue from theoretical discussion to policy concern,
- Pre-negotiational issue (1988- 1990) active involvement of governments to seriously take up the issue,
- Constitutional phase (1991- 1995) adoption and enforcement of the FCCC,
- Regulatory phase (1995- 2007): basically to implement provisions of the Kyoto Protocol by negotiating and elaborating the future climate regimes,

Between Rio (1992) and Kyoto (2005) several international protocols were adopted by global community (Montreal Protocol, Berlin Mandate, Johannesburg Summit) but it was in Kyoto (Japan) that a legally binding climate change framework was adopted by Western Countries except the United States which unilaterally refused to ratify the Protocol almost crippling it. In 2011 Canada withdrew from the Treaty further weakening it. In December 2009 another major Conference was held in Copenhagen. The Conference aimed to reduce emissions in half from 1990 levels by 2050. Serious differences between three high emitters–US, China and India- and developing countries particularly India and China, despite US president Obama's intense efforts, sunk those efforts. "International Climate politics have fettered like a mortally wounded animal" (http://www.spiegel.de/international/world/0.1518,druck-692861,oohtml). Copenhagen demonstrated not only the national- interest oriented narrow politics but also a lack of political will and vision. The next conference was held in Cancun Mexico in November 2010 with better results "achieving a fair bit of what is overhyped and acrimonious predecessor (at Copenhagen) had fluffed". (The Economist, December 10/2010:16) "The UN Climate process did quite well out of Cancun. The climate not so well". (The Economist 2010, December 18:22) Next major International Conference was convened in Paris in December 2015 and was considered as a watershed agreement on climate change. According to The New York Times (December 1/2015) besides 30,000 officials from 195 countries the event forged a partnership between prominent business leaders and philanthropists such as Microsoft's Bill Gates pledging substantial donations to reduce emissions, developing alternative energy sources, conservation, and aiding poor and low lying countries expected to be most affected by climate change. While the Paris Conference did not solve or even agreed on all aspects of climate change it did "bring countries together to undertake deeper cuts before 2030 and mobilize resources to implement stronger domestic reforms" (Oberoi 2017). It was realized that "aggressive climate targets are still within the reach if countries enact a virtuous cycle of ever more aggressive climate action as outlined in the Paris Agreement". (Natural Resource Council, December 2015). The Conference emphasized the need for substantial investment of resources in energy technologies and identified three major challenges related to climate change.

- Protecting Antarctica,
- Reducing automobile emissions,
- Water shortages, and adopting better water management technologies.

The conference also provided a time table to achieve these goals. The Paris Agreement realized several limitations in this context. Even if nations met initial Paris pledges to reduce carbon emissions a 3.5C temperature would rise by 2100, and could submerge coastal cities and drive over half of all species to extinction (Time, December 28- January 4/2016) (Figure 10)

As of February 2020, all UNFCCC members have signed the agreement, 189 have become party to it, and the only significant emitters which are not parties are Iran and Turkey. ("Paris Agreement". United Nations Treaty Collection. 8

July 2016)

Overall with Paris Agreement "a great tide had turned, finally the world stood united against the central environmental challenge of our time, committed to cutting the carbon pollution that is driving climate change. This Agreement set us on a course of verifiable gains we can build on over time... climate change has met its match in the collective will of a united world". (Rhea Suh, President, Natural Resource Defense Council December 2015). The Chinese President Xi observed that climate change is a "huge challenge". And it is very important for China and the United States to be firmly committed to the right direction of building a new model of major country relations". (The New York Times, December 1/2015)

Following the Paris Conference more than 170 countries reached a legally binding accord in Kigali in October 2016 to meet climate change by cutting the world-wide use of powerful planet warming chemical used in air- conditioners and refrigerators ... American president Obama called the deal "an ambitions and far reaching solution to this looming crisis". The Kigali Amendment to Paris Agreement added momentum to a series of new global climate change agreements. Negotiations from India and some of the world's hottest poor countries pushed back hard. As a result the final deal divided the world economy into three tracks:

- The richest countries including the United States, and the European Union countries to freeze the production and consumption of HFCs by 2018; reducing them to about 15 percent of 2012 levels by 2036;
- Much of the rest of the world including China, Brazil, all of Africa, India, Pakistan, Iran, and Kuwait will freeze HFC use by 2024 reducing it to 20 percent of2021 levels by 2045,
- A small group of world's hottest countries- India, Pakistan, Iran, Saudi Arabia and Kuwait- will have the most lenient schedule freezing HFC use by 2028 and reducing it to about 15 percent of 2025 levels by 2047 (The New York Times, October 16/2016). The Indian Prime Minister Modi observed that "Climate change is not our (poor countries) making. It is the result of global warming that came from the prosperity and progress of an industrial age powered by fossil fuel".

As discussed above issues related to climate change/ rise of temperatures/ global warming is existentialist challenge facing humanity. Answers and solutions to these challenges is not easy. There is no silver bullet. "No human intervention could prevent natural catastrophes, although a better early warning system, better popular education about evasive strategies, and faster and more effective emergency responses would surely mitigate the consequences"

(Parker 2014:689). Scientists and policy makers have struggled, disagreed and even denied the impact of climate change on Earth; but denying the scientific facts based events is not the answer. Many suggestions have been made to minimize the consequences of climate change. Following is a brief analysis of most discussed suggestions.

What is to be done? Policy prescriptions:

It has been observed that "Carbon dioxide concentration is largely irreversible for 1000 years after emissions stop" (Solomon et al. 2009:1750; Dessler and Parson 2006: 91-127) and several others (O Brian et al. 2000: 221-232; Victor et al. 2009: 64-76) have suggested following measures to at least slow down the disastrous consequences of climate change on Earth's eco- system.

Adaptation: The Fifth Assessment Report of IPCC defines adaptation as the "process of adjustment to actual or expected climate and its effects.. to change the fundamental attributes of systems in response to actual or expected climate and its effects, often at a scale and ambition greater than incremental activities. It includes changes in livelihoods from cropping to livestocks or by migrating to take up a livelihood elsewhere and also changes in our perceptions and paradigms about the nature of climate change, adaptation and their relationship to other natural and human systems. (Noble and Huq 2014:5). Excessive consumerism, wasteful exploitation of natural resources, unsustainable life styles, our politics, economic and cultural world view have put tremendous demands and pressures of Mother Nature carrying capacity. Humans need to change their promiscuous life styles. To quote an African proverb "if you cannot turn the wind, so turn the sail".

Mitigation: Dessler and Parson (2006: 96) observe that "... adapting to the impacts of climate change is not by itself an adequate response, it is also necessary to target the causes of climate change through mitigation measures, thereby slowing the impacts that we have to adapt it". To mitigate climate change, spare nature and address global poverty requires nothing less than "intensifying many human activities- particularly farming energy extraction, forestry and settlement so that they use less land and interfere less with the natural world" (Time, April 15/2015). The social and economic cost to mitigation will be painful in the short term, and therefore, there is a tension between adaptation and mitigation to some, the former implies a disregard of the later, as if society were giving up on trying to reduce green- house gas (GHC) emissions"(Henson 2011: 32). The Eco-modernists believe that economic development is an indispensable precondition to preserving the environment, and achieving it will require dropping the goal of sustainable development supposedly in harmonious interaction with nature, and replacing it with a strategy to shrink humanity's footprints by using nature intensively (Time, April 15/2015)

Integrated Assessment: Integrated assessment model advocate a combined approach- adaptation and mitigation- to face the climate change adversities. According to Dessler and Parson (2006:121) "Integrated Assessment models represent the climate system, the socio- economic factors that drive emissions, the impacts of climate change and potential mitigation and adaptation responses in a consistent quantitative framework. While highly simplified these models can be used to simulate the effects of different mitigation and adaptation strategies, to calculate the costs and benefits of alternative scenarios and policies".

Geo-engineering: Scientists and policy-makers have suggested a 'most cost- effective technological approach to meet climate changes by 'manipulating' or toying with climate "to offset the impact of increased greenhouse effects in the eco- sphere of the Earth by 'geo- engineering' using technology which will include diverse methods to cool the Earth. It would involve solar radiation management by seeding clouds, installing grant mirrors in the desert, injecting sulfates into the stratosphere, removing carbon directly from the atmosphere and burying it into oceans (Spector 2012: 99). The supporters of geo-engineering approach feel that "it is time to take geo-engineering out of the closet" (Victor et al. 2009: 74-76), however there are serious doubts about the practicality of the approach, and some people consider it not only very costly but plausible to absurd". It has been observed that "decoupling of human welfare from environmental impacts will require a sustained commitment to technological progress and the continuing evolution of social, economic and political institutions alongside those changes" says the Eco-Modernist Manifesto" (The New York Times, April 15/2015). Also there is a need to rethink development. Development paradigm must change- not based on per capita income and other indicators of GDP. Climate change must be de-ideological. A change in government must not derail it. Hopefully, Earth should not be doomed, and climate should survive Trump (The New York Times, November 30/2016; The Economist, November 19/2016:56)

Notes: 1. The Ozone is measured in Dobson units "One Dobson unit is the number of molecules of Ozone that is required to create a layer of pure Ozone 0.01 millimeters thick at a temperature of 0 degrees Celsius and a pressure of 1 atmosphere". (Ozone whole watch: What is a Dobson unit, March 15/2012)

REFERENCES

10. Assad Eduardo et al. 2013 '*Climate Change: Challenges for Brazil*', in Vajpeyi D. (ed.) 2013.

11. Astill, James 2012 '*The Vanishing North*', The Economist, June 16-22.

12. Bodansky Daniel and Rajmani Lavanya 2015 '*Evolution and governance Architecture of the Climate Change Regime*', Detler Sprinz et al. (eds.) 'International Relations and Global Climate Change: New Perspectives 2016'(forthcoming) 2016 SSRN: http://eern.com/abstract=2168859

13. Caldwell, Lynton K. 1984 '*International Environment Policy*', Durham, N.C., Duke University Policy Studies.

14. Climate Change 2007 '*The Physical Science Basis-summary for Policymakers, Fourth Assessment Report of IPCC on Climate Change*' www.ipcc/pdf/assessment_report/ar4- wgl-spm,pdf.

15. Crist, Meehan 2018 " *Besides I will be Dead* ", a review article of Goodell, Jeff 2017 " The Water will come". Boston: little Brown,in London Review of Books, 22 February 2018 :12-13.

16. Deaton Angus 2013 '*The Great Escape: Health, Wealth and the Origins of Inequality*', Princeton: Princeton University Press.

17. Dessler, Andrew, and Edward A. Parson 2006 '*The Science and Politics of Global Climate Change: A Guide to the Debate*', Cambridge, U.K. Cambridge University Press.

18. Dwivedi, O.P. 2008 '*The Well-being of Nations Reflections on integrating the Human and Eco-system well-being*', in Dhirendra Vajpeyi and Renu Khator (eds) 2008 'Globalization, Governance and Terchnology Challenges and Alternatives', New Delhi: Deep and Deep: 60-75.

19. Easterling, D.R et al. 2000 (a) '*Climate Extremes Observations, Modeling and Impacts*, Science 289, no 5487; Clivac.ucsb.edu/classes/tropical meteorology/classes- ucsb/climate_change/articles_climate change/articles_climate_change_tropics/Easterling et al._2000%extreme%20 events.pdf'

20. Easterling, D.R. et al. 2000 (b) "*Observed variability and Trends in Extreme Climate Events: A Brief Review*", Bulletin of the American Meterological society, 81, no. 3, gol.cniten/uploadfile/file/2000/2012010714282449.pdf

21. Elmer- Dewitt, Philip 1992 'Rich is Poor', Time, June.

22. '*Environmental Policies in the Third World: A Comparative Analysis*', Westpost, Con: Greenwood Press, 1-21.

23. Fang Ting Cheng 2013 '*Climate Change and its Impacts: Domestic and International Response of Japan*' in Vajpeyi, D. (ed.0 2013)

24. Flannery, Timothy 2006 '*The Weather Makers: The History and Future impact of Climate Change*', London: Allen Lane.
25. Gaby, Keith 2011 '*Climate 411, New Report on Climate Change says wilder Weather is headed our way*', *Climate Central*, November 18.
26. Gardiner, S.M 2011 '*A Perfect Normal Storm- The Ethical Tragedy of Climate Change*' New York: Oxford University Press.
27. Giannini A et. al. 2003 '*Oceanic forcing of Sahel rainfall on Inter-annual to Inter- decadal Timescales*', Science 302: 1027-30.
28. Grigoryev, Leonid et al. 2013 '*Domestic debates on Climate Change in Russia*', in Vajpeyi, D. (ed.) 2013.
29. Grubb, Michael 1995 '*Seeking fair Weather: Ethics and International Debate on Climate Change*', International Affairs, 71(3) 463-496.
30. Hartsgaard Mark 2011 '*Hot-living through the next fifty years on Earth*', New York: Mifflin Harcourt.
31. Henk, Dan 2005 '*Human Security: Relevance and Implications*', Parameters, summer; 91-96.
32. Henson, Robert 2011 '*The Rough Guide to Climate Change*', London: Rough Guides Limited.
33. Hoffman, A.J. 2012 '*Climate Science as Culture War*', Stanford Social Innovation Review, Fall, 30-37.
34. Huber, Daniel G., and Jay Gulledge 2011 '*Extreme Weather and Climate Change: Understanding the link and managing the risk*', Center for Climate and Energy Solutions, www.c2es.orgwww.c2es.org/docuploads/white-paper-extreme-weather- climate-change-understanding-link-managing-risk.pdf.
35. Jian Li 2013 '*China and Climate Change: Environmental Impacts, Human Security and Mitigation policies and actions*', in Vajpeyi D. (ed.) 2013
36. Johnson, Craig 2013 '*Accommodating Climate Refugees: Models of Sovereignty and Security in the International Climate Regime*', in Vajpeyi, D (ed.) 2013.
37. Karl, Thomas 2008 '*Weather and Climate extremes in a changing climate regions of focus: north America, Hawaii, Caribbean and U.S. Pacific Islands, Washington D.C.: U.S. Climate Change Science Program*', purl-access.gpo.gov/GPO/LPS11076
38. Koenig, Dieter 1995 '*Sustainable Development: Linking Global Environmental change to Technology Cooperation*', in O.P. Dwivedi, and Dhirendra Vajpeyi (eds.).
39. Lovelock, James 1987 '*Geo-Physiology: A new look a Earth Science*', 'The Geo- Physiology of Amazonia: Vegetation and Climate Interactions' (ed.) Robert E. Dickenson, New York: John Wiley Meadows, Donella et al. 1972 '*The Limits to Growth*', New York: Universe Books.

40. MacNeill, J, et al. 1991 '*Beyond Interdependence*', New York: Oxford University Press'.
41. Mc Donald, Mary 1998 '*Agenda for Sustainability, Environment and Development in the twenty first century'*, New York: Routledge.
42. Mendelsohn, Robert et al. 2006 '*The Distributional impact of climate change on Rich and poor Countries*', Environmental and Development Economics, 11.
43. Miller, J. Hills 2016 '*Thinking Literature Across Continents*' in Ghosh, Ranjan and J. Hills Miller, London: Duke University Press.
44. Mishra, R.K. et al. 2013 '*Climate Change in Bangladesh and Nepal'* in Vajpeyi, D. (ed.) 2013.
45. Myers, norman 1993 '*Ultimate Security: The Environmental Basis of Political Stability*', New York; W.W. Norton.
46. Nef Jorge 2008 '*Political Economy of Globalization*', in Vajpeyi and Khator (eds.) 2008.
47. Noble, Ian R, and Saleem-ul-Huq 2014 '*Adaptation needs and options', in 'Climate Change 2014; Impacts, Adaptation and Vulnerability*' IPCC 5th Assessment Report, Working Group II.
48. Nordhans, William 2012 '*Why the Global warming skeptics are wrong*', The New York, Review of books, March 22, 32-34.
49. Oberoi, Roopinder 2017 '*Sustainable Development and Globalization*', in Dhirendra Vajpeyi, and Roopinder Oberoi 'Globalization Reappraised; An Oracle or a Talisman (forthcoming)', New York: Lexington Books.
50. Oberoi roopinder et al 2013 '*Climate Change: Vulnerabilities and Human Insecurities- Understanding the impact on India*' in Vajpeyi, D. (ed.) 2013.
51. O' Brian, Karen L. and Robin Leichenko 2000 '*Double Exposure: Assessing the impacts of Climate Change within the context of economic Globalization*', Global Environmental Change, 10, 221.
52. "Paris Agreement". United Nations Treaty Collection. 8 July 2016. Archived from the original on 21 August 2016. Retrieved 27 July 2016. (https://treaties.un.org/pages/ViewDetails.aspx?src=TREATY&MTdsg_no=XXVII-7- d&chapter=27&clang=_en).
53. Paris, Roland 2001 '*Human security: A Paradigm Shift or Hot Air*', International Security, Autumn: 87-108.
54. Parker, Geoffrey 2014 '*Global Crisis: War, Climate Change and Catastrophe in the seventeenth century*' New Haven: Yale University Press.
55. Parry, M and t. Carter 1998 '*Climate Impact and Adaptation Assessment*', London: Earthscan.

56. Randall, David A. and Richard A. Wood et al. 2007 '*Climate Models and their Evaluation*', Climate Change 2007: The Physical Science bases contribution of Working Group I to the Fourth Assessment Report of the Intergovernmental Panel on Climate Change, Cambridge: Cambridge University Press.
57. Reifen C and R. Toumi 2009 '*Climate Projections: Past Performance no guarantee of future skill? Geophysical Research Letters 36*', 10;1029GLO38082.
58. Rittel, Horst and Melvin Weber 1973 '*Dilemmas in a General Theory of Planning*', Policy Sciences, http://www.jastor.org/stable/4531523
59. Singer, S, Fred 2012 '*The Climate Contrarians*' The New York Review of Books, August 16.
60. Solomon, Susan et.al. 2009 '*Proceedings of the National Academy of Sciences*', 106; 1704-1769.
61. Spector, Michael 2012 '*The Climate-Fixers- Is there a technological solution to Global Warming*', The New Yorker, May 14: 96-103.
62. Tanik, Laura 2013 '*Health Human Security and Climate Change*', in Vajpeyi, D. (ed.) 2013.
63. Vajpeyi, Dhirendra 2013' (ed.) '*Climate Change, Sustainable Development, and Human Security: A Comparative analysis*', New York: Lexington Books.
64. Vajpeyi, Dhirendra (ed.) 1995 '*External factors influencing Environmental Policymaking: Role of Multilateral Development Aid Agencies*', Environmental Policies in the Third World, Westport, CT: Greenwood.
65. Vajpeyi, Dhirendra (ed.)2001 '*Deforestation, Environment and Sustainable Development: A Comparative Analysis*', London: Praeger.
66. Van Aalst, Marten K 2006 '*The Impacts of Climate Change on the Risk of Natural Disasters*', Disasters.
67. Victor, David et al. 2009 '*Green engineering option- A last resort against Global Warming*', Foreign Affairs, March-April: 64-76.
68. Welsh, Jennifer 2009 '*Implementing the Responsibility to Protect*', UK; Oxford, Institute for Ethics, Law and Armed Conflicts.
69. White, Clifton, and Rebecca McGuiri 2013 '*Global Warming and Climate Change Extremes*', in Dhirendra Vajpeyi (ed.): 27-49.
70. Williams, Paul 2013 '*Climate Change its effects, and the Political Economy of Adaptation and Mitigation: Turkey and the Eastern Mediterranean Region*', in Vajeyi, D. (ed.) 2013.

71. Woolings, T 2010 '*Dynamical Influences on European Climate: An Uncertain Future, Philosophical Transactions of the Royal Society A368*'.

72. Zeng, N 2003 '*Drought in the Sahel*', Science, 302: 999-1000, www.sciencemag.org/content/302/5647/999.short

Chapter 2

Impact of Urban Development on Environment

Vishwmaitri Shekhawat and Ashok Kumar Gupta

ABSTRACT

Globally, over 50% of the population lives in urban areas today, and this trend is expected to continue - by 2045, the number of people living in urban areas will increase by 1.5 times to 6 billion. Urban areas consume close to 2/3 of the world's energy and account for more than 70% of global greenhouse gas emission. As cities develop, their exposure to climate and disaster risk also increases. In recent decades, the urban environment of developing countries has been subject to increasing air, water and land pollution. Although urban development is perceived as necessary for a sustainable economy, but uncontrolled and rapid urban development causes various problems. Considering this in view the present paper seeks to analyse the relationship between the urban development process and environment. I would also analyse the formulation of effective policies for a healthy urban environment and to gain a sustainable urban development.

Keywords: urban development, urban environment, pollution, sustainable environment

INTRODUCTION

There is a general belief that the economic, demographic, and technological developments in urban areas increasingly impose strain on the urban environment. According to Clark (1982) urban development is the process of emergence of a world dominated by cities and urban values and the main process of urban development are urban growth and urbanisation. Urban growth is a spatial and demographic process and refers to the increased importance of towns and cities as concentration of population within a particular economy or society, while process which refers to the change of behaviour and social relationships which occurs in society as a result of people living in town and cities.Globally, over 50% of the population lives in urban areas today, and the trend is expected to continue- by 2045, the number of people living in urban areas will increased by 1.5 times to 6 billion, adding 2 billion more urban residents (The World Bank 2016). This rapid growth of urbanisation coupled with fast pace of industrialisation is responsible for serious disturbance in the ecological balance and as a result we are experiencing urban environmental hazards and rapid deterioration in the quality of urban environment. Modern cities are growing so rapidly and this rapid advancement of urban development has given rise to manifold environmental pollution and problems. In the present century, environment degradation has emerged as a major global problem due to industrialisation and urban growth. Although urban development is perceived as necessary for a sustainable economy and no doubt has a positive impact on income level, employment, production, infrastructure, and education but uncontrolled and rapid urban development causes various problems including inadequate sanitation and waste disposal facilities, inadequate water supply, slums, worsening water quality, excessive air pollution, noise, dust and heat. According to the Global Change Program from university of Michigan, urban development wipes out animal ecosystem, causes water and air pollution, depletes natural resources. Environmental hazards of urban development affect households, neighbourhoods, workplaces, municipalities, regions and cities. The Global 2000 Report to the President (1980) concluded that if present trends continue, the globe will become more crowded, more polluted, ecologically less stable and more vulnerable to description than the world we live in today. And today we are seeing it becoming true.

A brief discussion is presented here of degradation of environment in Indian cities due to urban development.

URBAN DEVELOPMENT IN INDIA

Urban India is growing at an unprecedented rate. India is urbanizing rapidly by crossing 31.2% level of urbanization. In terms of number about 377 million people live in 7935 Indian cities/towns (census 2011). The number of towns significantly increased from 5161 in 2001 to 7935 in 2011. There is high inequality in the distribution of urban population by size class categories. About 70% of the total population (265 million) live in 468 towns. Not only the number of people and number of towns but also the urban land area is increasing rapidly. The cities in India show peripheral expansion where small municipalities and large villages merge with the core city and become part of the large metropolitan area.

Table 1: Urban Population and No. of Urban Agglomerations in India (1961-1971)

Census year	*No. of urban agglo-merations*	*Total urban population*	*Urban population in %*	*Annual growth rate % of urban population*
1961	2363	78936603	17.97	1.71
1971	2590	109113977	19.91	2.63
1981	3378	159462547	23.33	0.66
1991	3768	217177625	25.72	1.65
2001	5161	285354954	27.78	1.54
2011	7935	377105760	31.16	1.10

Table 2: Growth of Urban Population by City Size (percent per annum)

Types of cities size and class	*1971-1981*	*1981-1991*	*1991-2001*	*2001-2011*
Cities	4.4	3.7	3.5	2.7
Metropolitan cities	4.2	4.9	4.2	3.5
Class 1	2.7	2.4	1.5	1.6
Class 2	4.1	2.8	1.6	1.6
Class 3	2.4	3.0	1.9	1.6
Class 4+	1.9	1.3	1.0	1.6

Source–various reports of Census of India

McKinsey's report on India's Urban Awakening (2010) predicts that about 590 million people will be living in urban areas by 2030 in India and equates it to twice the total population of U.S.

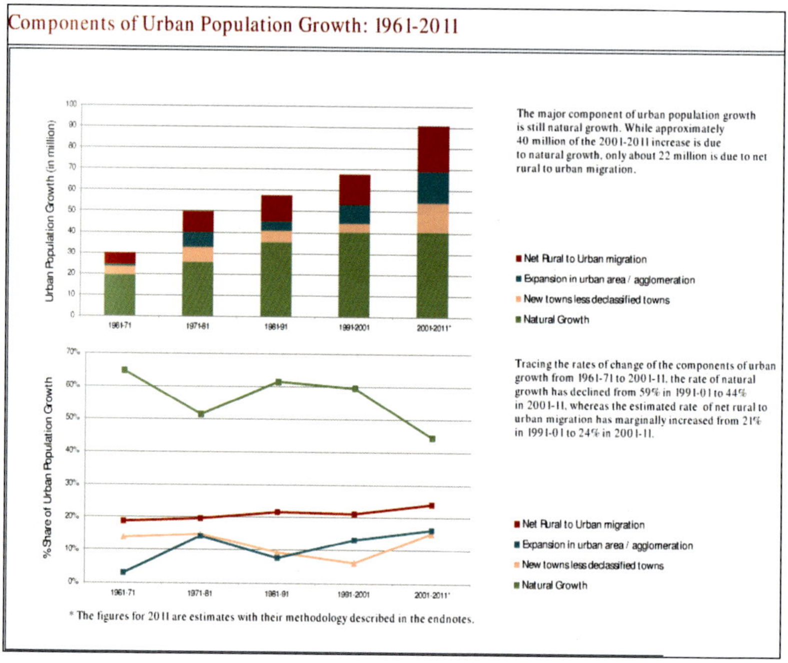

Figure 1: Components of Urban Population 1961-2011

Source: IIHS Analysis

To accommodate this population, the report estimates that about 700-900 square million metres of residential and commercial space would have to be like creating a new Chicago city every year besides the 2.5 billion square metres of roads which have to be built and 7400 k.m. of metro and subways to be constructed, estimated to be 20 times the capacity added in the last ten years. The report of High Powered committee on Indian urban infrastructure and service (2011) says the urban area's share in the national GDP is bound to increase from 62- 63% in 2009-10 to about 75% in 2030. (Figure 1)

Impact of Urban Development on Environment

[A] Creation of Heat Island

Higher temperature in centre urban cities than that in suburbs is called Urban Heat Island (UHI). Rapid urban growth and high population density coupled with densely built up area and lack of greenery in the cities creates UHI effect. This effect is mainly contributed by built-up surface which are composed of non-reflective materials. It absorbs the incident radiation and releases it as heat. In India with the development of cities, built-up area is increasing rapidly, which is making Indian cities vulnerable to the impact of UHI.

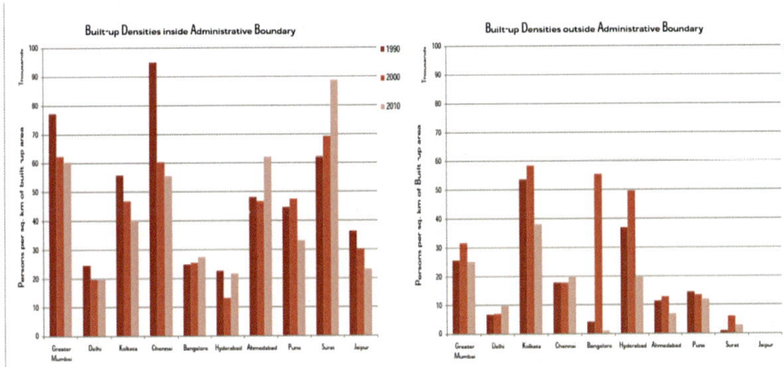

Built-up density, estimated as population over built-up, is decreasing for most of the core areas of the ten largest cities. The evolution of density outside the urban local body boundaries varies more, but density is lower than that in the city cores.

Figure 2

Source: H.S. Sudhira (2011) and IIHS Analysis

[B] Poor Air Quality

The quality of air in cities is a major environmental concern. It indirectly affects climatic conditions and directly affects the health of community. Increase in industrial activities, economic activities, population and vehicular population etc. have led to a rapid increase in environmental problems, one of them being air pollution. Rising anthropogenic air pollution is a direct threat to our environment and is accelerated by growing urban population with increasing dependency on coal and petroleum as energy sources. The impact of pollutants on climate is increasingly visible as changes in hydrological cycle, temperature and precipitation. The highest concentration of air pollutants are found in the urban areas (NO_2, SO_2, RSPM) due to congestion of traffic, poor quality of fuel, inefficient engines and bad maintenance of motor vehicles (Central Pollution Control Board, CPCB). Since 2011, at least 94 Indian cities have not met national air quality standards. According to World Health Organization(WHO), 37 cities from India feature in the top 100 world cities with the worst PM10 pollution, and the cities of Delhi, Raipur, Gwalior and Lucknow are listed in the top 10 (WHO,2014). A similar assessment by WHO, in 2011, listed 27 cities in the top 100.

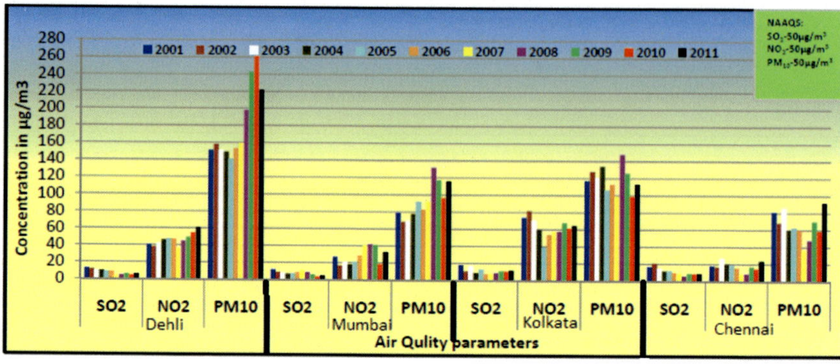

Figure 3: Air Quality Trends in Four Mega Cities

Source :CPCB 2012-2013

As cities are increasing in size and population, there is a steady demand for motorized vehicles in both personal and public transport sector. This puts substantial pressure on the city's infrastructure and environment, particularly since most Indian cities have mixed land use. The most commonly identified sources of air pollutants in urban India are vehicles, manufacturing and electricity generation industries, construction activities, road dust and waste burning; which are obvious result of development of urban area.

[C] *Degradation of Water Resources*

Urban development affects the quality and quantity of the water resources. The construction of roads, buildings and increased built up area affects and changes the natural recharge patterns. Water requirement from urban communities and their economic activities and commercial enterprises become increasingly demanding, which is reducing ground water level.

The water quality has also degraded with time due to urbanization that ultimately leads to increased sedimentation there by also increasing the pollutant in run off. According to annual report of CPCB 2013-2014 the organic and bacterial contamination continued to be critical, primarily due to increasing discharge of untreated sewage generated from urban centres which the municipal bodies are unable to handle. Use of subsurface for urban engineering-pipes, sewers, tunnels for roads and metro systems, deep basements and foundations- and for provision of building materials can also affect shallow ground water beneath cities more indirectly. The benefits of these activities are usually apparent at the outset, but the environmental impacts and their associated costs may not be appreciated until much later.

[D] *Polluted Land Resources*

The accumulation of different form of solid waste is the major aspect of land pollution in urban area. Indian cities produce roughly 57-68 MT of solid waste per year and the safe way of their disposal is becoming increasingly difficult. Many of the urban cities in India do not have either proper drainage system or following proper method of disposal. Heaps of solid waste can be seen scattered in the cities. The improper solid waste disposal leads to urban land pollution which in turn causes public health hazards, depressed property values, inefficient use of scarce urban land and affects aesthetic sensibilities. In addition to unbearable stink, this is hazardous to the health of the residents and may lead to the spread of many diseases, lack of sufficient number of public toilets and latrines has added to problem of sanitation in the cities.

Table 3: Municipal Solid Waste Generation in Some Cities of India (tonnes per day)

Name of city	*1999-2000*	*2004-2005*	*2010-2011*
Ahmedabad	1683	1302	2300
Bangalore	2000	1669	3700
Chennai	3214	3036	4500
Delhi	4000	5922	6800
Hyderabad	1566	2187	4200
Kanpur	1200	1100	1600
Kolkata	3692	2653	3670
Mumbai	5355	5320	650
Pune	700	1175	1300
Surat	900	1000	1200

Source- MSW Report 2011-12

ENVIRONMENTAL PLANNING FOR SUSTAINABLE URBAN DEVELOPMENT

The World Bank notes that the negative effects of urban develop-ment can be prevented through provision of good sanitation, proper drainage, protection of environmentally sensitive lands and proper waste management. Urban environmental planning and management should come to centre stage of planning philosophy and development strategy. To make urban systems sustainably developed careful environment planning and management are necessity of today's urban world. Environmental aspects are to be inducted into each of the development activities at the planning stage itself and are to be well co-ordinated and balanced. At present the environmental aspects are not usually considered while preparing Master Plans and the process is skewed towards development need. Therefore there is a dire need for developing spatial planning tools & techniques with thrust in environmental reality. Integrated spatial environmental planning based on an assessment of existing environmental profiles as well as potential assimilative capacity could help environmentally acceptable development. The policies for urban environmental management and sustainable development should include:-

- Capacity building at local level
- Re-orienting the public policy
- Re-orienting the legal and administrative framework
- Development and disseminating urban environment database
- Integrated approach to environmental management
- Public consultation and participation
- Integrating environmental considerations in urban development plans

CONCLUSION

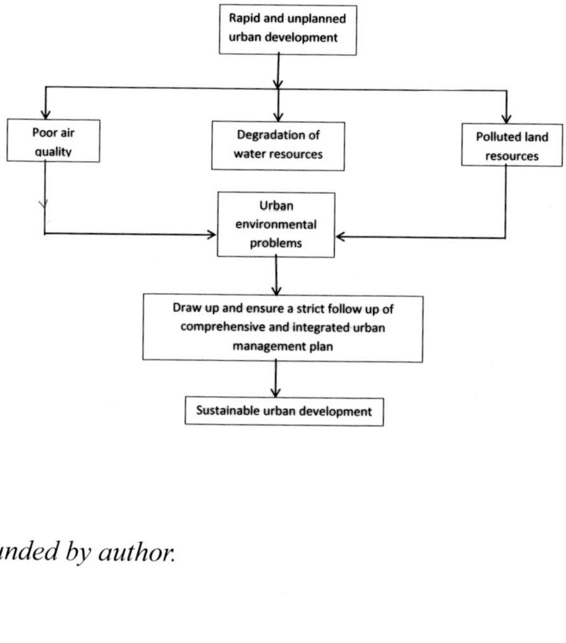

Figure 4

Source: Compounded by author.

Due to rapidly growing populations the demand for basic infrastructure and services in Indian cities is increasing. These unmet demands adversely affect the quality of urban life as well as the process of sustainable development. Environment has deteriorated due to unplanned growth of cities and has led to deterioration of quality of air water and land.

In order to manage urban environmental problems and to gain sustainable urban development, formulation of effective policies is necessary. To find policy measures to mitigate these problems, a multidisciplinary approach is needed. Not only governments but also the society has to play an important role to take action now, to shape the future of our development, to create opportunities for all.

REFERENCES

1. Central pollution control board annual report (2011-12)
2. Central pollution control board annual report (2013-14)
3. Clark D. (1982), Urban geography: An Introductory Guide. The Johns Hopkins University press, Baltimore.
4. Guttikund S.K., Goel R., Pant P. (2013), Nature of Air Pollution, emission sources and management in the Indian Cities. *Atmospheric Environment,* vol. 95:501-510
5. Nyambod E.M.(2009), Environmental Consequences of Rapid Urbanisation Bamenda City, Cameroon. *Journal of Environmental Protection*, vol. 1: 15-23
6. Status Report on Municipal Solid Waste Management (2011-12), Central Pollution Control Board, Delhi.
7. United Nations, Department of Economic and Social Affairs, Population Division (2016), The World's Cities in 2016 Data Book.
8. United Nations, Department of Economic and Social Affairs, Population Division (2014), World Urbanization Prospects: The 2014 Revision, Highlights.
9. Urban India 2011: Evidence (2011), Indian Institute for Human Settlement.
10. William R.A. (2000), Environmental Planning for Sustainable Urban Development. Paper presented in Caribbean water and wastewater association 9th annual conference & exhibition at Chaguaramas, Trinidad.

Chapter 3

Do Women in Government Champion Development Friendly Regulations?

Brian Warby and Silvia Pellegrino

ABSTRACT

Although microfinance has not proven to be the cure-all for global poverty, the evidence and rhetoric was overwhelmingly positive for many years, and it played a significant role in reducing the impact of loan sharks and unregulated money lenders that had represented the only options for the rural poor in LDCs (Dasgupta and Rao 2003; Kumar 2013). So why did some governments adopt policies that were friendly to microfinance while others did not? We argue that female legislators fight for different kinds of policies than their male counterparts, and therefore, the gender composition of the country's legislative body is a key determinant in the microfinance business environment of a country. However, the evidence does not support this assertion. The conclusion offers some discussion of why that may be the case.

Keywords: Gender, WID, Microfinance, LDC, GAD

Some governments seem to operate more in the interests of their citizens than other governments. Some adopt policies that more effectively alleviate poverty, grow the economy, promote democracy and improve education, but others seem either unable or unwilling to do the same. For several years surrounding Mohammed Yunis winning the Nobel Prize for his work with the Grameen Bank, microfinance was considered by many people to be a groundbreaking poverty eradication tool. International institutions were promoting it, NGOs were adopting it and governments were shaping policies to facilitate it – that is to say, some governments were facilitating microfinance, while others were not.

Despite the microfinance industry having faced some major obstacles, it has recovered and remains a force in many developing countries, in particular India. The microfinance crisis in Andhra Pradesh in 2011-2012 caused the sector to contract and raised questions about its viability (Kaur and Dey 2013; Mader 2013). More recently, the 2016 demonetization of India punished the microfinance sector, reducing loan repayments and, therefore, loan disbursement (Rogers et al. 2018). In both cases the industry adapted and rallied. Today there are approximately 225 microfinance institutions operating in India, which serve millions of people every year.

Although microfinance has not proven to be the cure-all for global poverty, the evidence and rhetoric was overwhelmingly positive for many years, and it played a significant role in reducing the impact of loan sharks and unregulated money lenders that had represented the only options for the rural poor in LDCs (Dasgupta and Rao 2003; Kumar 2013). So why did some governments adopt policies that were friendly to microfinance while others did not? We argue that female legislators fight for different kinds of policies than their male counterparts, and therefore, the gender composition of the country's legislative body is a key determinant in the microfinance business environment of a country. However, the evidence does not support this assertion. The conclusion offers some discussion of why that may be the case.

THE APPEAL OF MICROFINANCE

There are a variety of different kinds of poverty interventions an LDC government could promote or adopt. Many of the interventions that get a lot of attention in the developed world are those that have some role for the developed countries. However, there are often a host of problems that can accompany those policies. Foreign aid has been the target of criticism for decades due to its variegated flaws (Easterly 2001, 2006) ranging from politically motivated donations that reward political subservience to big donors (Alesina and Dollar 2000; Morgenthau 1962), to its unpredictability for recipients (Kodama 2012). Aid exacerbates corruption (Fortin 2013), and can make fragile political circumstances more fraught (Chauvet and Guillamont 2004; Bräutigam and Knack 2004; Hubbard and Duggan 2009; Girod 2011). Foreign aid may be helpful, if managed correctly, but it comes with many pitfalls and leaves the LDC recipient dependent on donors.

International financial institutions, such as the IMF or World Bank have been involved in development and poverty alleviation. As international institutions, they may be free of some of the political problems we associate with foreign aid, but they have their own problems. Woods' (2001) critique laid bare some of the key problems that drive their ineffectiveness, including lack of transparency and accountability. Others have criticized the corruption and hypocrisy within these institutions (Weaver 2008), while still others have taken them to task for their failure to account for non-monetary concerns, such as environmental protections (Rich 1994). Perhaps worst of all were the misguided Washington Consensus policies they pushed struggling LDCs to adopt in past decades. Needless to say, LDC governments might be leery of these institutions.

LDC governments that genuinely want to promote inclusive economic growth might adopt policies that make the state more attractive for investment, both foreign and domestic, but courting international capital often requires changes that take considerable time (Z. Q. Wang and Swain 1995). Other changes may come at the detriment of domestic workers and residents (Li 2006; De Mooij and Ederveen 2003). Moreover, there are no guarantees that any particular policy change or set of changes will draw in much foreign direct investment (FDI) because FDI is driven by markets, which are complex and constantly changing (Dunning 1980, 2001)

If an LDC develops and implements its own poverty alleviation program, it is not subject to the whims of capricious foreign aid donors, nor the opaque and hypocritical policies of the World Bank. It does not require the upfront investment and risk associated with restructuring tax policies or other major elements of the economy to try and attract international capital either. Domestically funded and operated poverty interventions can be tailor made and do not depend on any outside actors' goodwill and effective implementation. Moreover, microfinance institutions are largely non-governmental, which means poverty alleviation and financial empowerment without having to spend a lot of government resources. Although microfinance has fallen out of favor to some degree over the last decade or so (Roodman 2012; Warby 2015; Bateman 2010), it was very popular and highly regarded by most people for a number of years, as evidenced by the United Nations naming 2007 the year of microfinance. So why didn't more countries try harder to foster a positive climate for microfinance?

GENDERED POLICYMAKING

One of the key features of many microfinance programs is that they often target women for loans (Imai, Arun, and Annim 2010; Navajas et al. 2000). There are a number of reasons that microlenders often target women to receive microfinancial services. On the normative side, many people believe that putting money into the hands of women will contribute to female empowerment. Women often have less of a voice in how to use the household's resources than their husbands, which is both a symptom of and contributes to their limited influence over their own household (Sen 2001), and their ability to invest in the things they value (Bonilla et al. 2017). They are usually in charge of child care and housekeeping, which is unpaid labor in itself, and it might keep them from participating in a formal workplace, especially in rural locations where technology advancements that commonly free up women's time are less available (Duflo 2011; Kabeer and Natali 2013)and negatively affect their performance and survival prospects. In this paper, we study how domestic supplier firms may adapt and continue to perform, as market liberalization progresses, through catch-up strategies aimed at integrating with the industry's global value chain. Putting money into women's hands might reverse those norms and improve gender equality. An increase in personal finances will also reduce women's vulnerability, increase control over fertility and family planning, boost their bargaining power, and remove women from dangerous environments due to an increase in opportunity (Duflo 2011; Kumar 2013). Others have argued that targeting women for development programs increases impact since they are more likely to allocate resources to health, nutrition, children's education and durable goods (Duflo 2012; Duvendack et al. 2011)

Women that borrow from micro-lending programs also have the ability to increase their output by investing in income-producing tools (Lott 2009). Microfinance institutions might additionally target women because of their tendency to be more compliant and disciplined and are, therefore, more affected by moral and societal pressure (D'espallier, Guerin, and Mersland 2013)

Microfinance institutions that target solely or mainly women are often non-profit NGOs with a focus on development. Despite the fact that only 47% of microfinance institutions loan solely to women, female borrowers make up 70% of microfinance customers (D'espallier, Guerin, and Mersland 2013)

Although focusing solely on women has proved to increase operational costs for a microfinance institution, women have proved to have a higher repayment rate. Higher costs are associated with a tendency to offer women smaller or group-based loans, therefore a shift in lending techniques and loan sizes might obviate the problem.

Women receive smaller loans, have higher interest rates, and face higher collaterals. Certain scholars point to discrimination as a cause for disparities (Treichel and Scott 2006)1979; Riding and Swift, 1990; Buttner and Rosen, 1992; Fabowale et al., 1995; Haines et al., 1999; Coleman, 2000, others to education gaps among men and women in developing countries(D'espallier, Guerin, and Mersland 2013)

Others argue that microfinance institutions were conceived with such characteristics because they targeted the poor in general, who do not have collateral, take smaller loans, and live in less accessible areas, but those characteristics have shifted towards women because they have proven to be better lenders (Lott 2009). Nonetheless, Armendariz and Morduch (2005) argue women are more responsible and less likely to borrow more than they can afford to repay. Indeed, microloans are better suited for women entrepreneurs, as they are less risky and have a quicker turnaround, compared to, for example, agricultural enterprises. Other studies, however, advocate for the use of microcredit for purposes other than entrepreneurship. Karlan et al (2016) and Prina (2015) argue that the use of microfinance for savings or necessity-based borrowing are equally as effective in lifting individuals out of poverty than entrepreneurial-based loans. Similarly, Dasgupta and Rao (2003) argue that microfinance aids women for labor and opportunity-related financial needs, through microcredit. But microfinance institutions do not emphasize microsaving, microleasing, and, more importantly, microinsurance programs for emergency-related needs.

Although microfinance institutions are generally non-governmental, they are nonetheless regulated by government policies. Therefore, the legislature is the natural starting point to understand why some governments adopt friendly policies while others do not. Both the positivist and normative perspectives on the value of targeting women as recipients of microloans may gain more traction among women than men. From the normative perspective, female legislators are more likely to have a better sense of what women care about and what they want or need (Celis 2006). This is one of the key points that the Women in Development (WID) literature makes. Women make up a significant portion of the poor and under-developed world, so policymakers should think about how to help them specifically, since the obstacles they face are not entirely the same as those that men tend to face (Connelly et al. 2000; Boserup 2011). In fact, having women in important positions, such as legislatures, is itself valuable in demonstrating to society that women are capable of rising to and filling important positions (York and Bell 2014)

The World Bank has suggested that empowering women and improving gender equality can have a range of benefits to society and the economy. Empowering women can lead to productivity gains by female laborers, higher human capital investments in children, since women are more likely to invest in their children's health and education, and more democratic societies through representative decision making when women have a political voice (World Bank 2012; Pande and Ford 2011).

WOMEN IN GOVERNMENT

Gender and Development (GAD) is a paradigm that evolved from the Women In Development perspective of the 1970s (Rathgeber 1990; Jaquette 2017).

GAD tries to understand the role that gender plays in development efforts from a holistic perspective. It considers the roots and norms of women's status in society in relation to that of men. It examines both productive and reproductive roles. It also acknowledges the potential contributions that men can make in advancing social justice and equality. GAD theorists often argue that politicians must consider these dynamics in order for meaningful changes to take effect (S. M. Rai 2011; Wanner and Wadham 2015).

Interestingly, the GAD perspective questions the similarities between women of different social and economic groups. There are certainly differences in the challenges facing wealthy women from developed countries and poor women in less-developed countries (Connelly et al. 2000; S. M. Rai 2011). The same may also be true at the societal level. Wealthy women who are likely to be in government likely have not had the same life experiences and difficulties as women in low-income households. Furthermore, Lovenduski and Norris (2003) suggest that female MPs must represent differences in backgrounds, personal finances, or ideologies relative to male MPs in order to push for qualitative change. This raises questions about how well powerful women are able to relate to and recognize the needs of poor women.

The question several scholars have already asked is whether women in important political positions, such as national legislative bodies, actually champion women's issues at higher rates than men. The answer, in short, is generally yes. Female legislators or parliamentarians (hereafter MPs) are more likely than their male counterparts to fight for gendered issues (Celis 2006; Wängnerud 2009). For example, US state legislatures with higher proportions of women are more likely to support legislation that we generally think of as protecting or advancing women's interests, such as parental leave protections (Williamson and Carnes 2013). Female MPs in Belgium were found to spend more time and energy advancing women's interests and they addressed the needs of some groups of women that had never adequately been addressed by male MPs (Celis 2006).

Others have come to similar conclusions, but with caveats, particularly that there is some threshold of women in parliament below which they are unlikely to be able to do much on women's issues because they won't be able to make headway and may be pushed out for not conforming to mainstream issues (Celis 2006; Reingold 2000; Tinker 2004) Others come to the opposite conclusion, that women in politics can shift the debate without having a critical mass (Ayata and Tütüncü 2008). Regardless, having more women in government can be a force for change. Most governments around the world have relatively low ratios of women to men. Even countries that are often thought of as being rather progressive on gender issues, such as the Scandinavian countries, have yet to achieve gender parity in their national parliaments as of 2017 (Schwab et al. 2017). OECD member countries have yet to collectively break the 30% threshold. While some countries have close to gender-balanced legislatures, LDCs as a group average just over 20% female MPs. Even when women get into public office, they are usually put in charge of what UN Women calls "soft issue" portfolios, meaning cabinets related to social affairs, family, children, youth, elderly, and disabled, with a recent shift toward environment and energy-related cabinets. In 2019, 22% of women ministers hold "soft issue" portfolios while only 1.5% of them hold the economy/development portfolio ("Guidelines for the Elimination of Sexism, Harassment and Violence against Women in Parliament" 2019).

Interestingly, India has remarkably low female representation in national government. The first bill to reserve seats for women in the Lok Sabha (national parliament) was first introduced in 1996. The bill has repeatedly failed to get the support it needs for adoption (Randall 2006). The Inter-Parliamentary Union places India at 149th of 193 countries on the percentage of women in the lower house, with just 11.8% women. Regional parties have done much better, achieving as much as 40% female candidates in some 2019 elections (Srivastava 2019). Regardless, studies have shown that female MPs' platforms in India vary greatly based on which caste they belong to. Female politicians of a higher caste invest less in education and more in policies that seem to perpetuate inequality while low caste MPs focus on legislation that advance women's autonomy (Clots-Figueras 2012). India is among the countries that might benefit a great deal from advancing gender equality in terms of political representation and beyond (Turnbull 2019).

Many microfinance programs have overtly gendered components. Having more women in parliament may make it easier for female MPs to champion policies that promote microfinance and are overtly gendered, both because women MPs tend to be more likely to write and sponsor legislation on women's issues and because having larger numbers of women means that there is a better chance of advancing the conversation and implementing solutions. GAD would suggest that having more women MPs should make a government more likely to adopt microfinance-friendly policies.

Hypothesis 1: A higher proportion of women in national parliaments will make a country more likely to adopt microfinance-friendly policies

As mentioned above, GAD looks beyond sex to include factors such as class and race (Everett and Charlton 2014). That is to say that a person's identity, and, therefore, her life experiences and the lens through which she sees the world, is a function of more than just sex or gender. While there is some overlap, a poor woman living in a rural village in northern Mexico may face starkly different obstacles than a middle-class woman just across the US border (S. Rai 2011).

Few people would argue with this assertion, but we might go further to suggest that well-educated, wealthier women, even in LDCs, may face glaringly different structural barriers than their poor compatriots.

A more nuanced examination of the literature on women in government fighting for women's issues suggests that men who are sensitive to gendered problems may contribute to the critical mass necessary to make headway on women's issues (Htun and Jones 2002; V. Wang 2013; Williamson and Carnes 2013). In the constructivist vein, wide acceptance of the idea that there is value in gendered policymaking may be more important for whether MPs support both the adoption and the continuation of microfinance than their own gender.

Hypothesis 2: Acceptance of the value of gendered policymaking will make a country more likely to adopt microfinance-friendly policies

The next section lays the framework for and executes empirical tests of these hypotheses.

EMPIRICAL TESTS

The empirical tests rely on data from the Economist Intelligence Unit (EIU) for the dependent variable. The EIU began reporting on the microfinance environment in 2007. The first two years only included data for Latin America and the Caribbean, but since 2009 it has included 55 countries. The score has a theoretical maximum of 100 and minimum of zero. The annual reports describe the methodology the EIU uses to develop the measure. In brief, they draw from a range of information sources including personal interviews, scholarly studies and government and industry reports ("Global Microscope on the Microfinance Business Environment" 2012). The likelihood of a country adopting microfinance-friendly policies is captured by the microfinance environment score.

Data on women in parliament comes from the Inter-Parliamentary Union database, and gender quotas are taken from the Varieties of Democracy data (Coppedge et al. 2017; Pemstein et al. 2017). The polity variable is polity2 from the Polity data set (Marshall and Gurr 2014). The Gender Inequality Index comes from the United Nations' *Human Development Report*. Economic data is from the *World Development Indicators*.

Table 1: Summary Statistics

Variable	*Obs*	*Mean*	*Std. Dev.*	*Min*	*Max*
Regulatory Env.	304	49.97796	16.92788	12.5	100

%women MPs	2,889	13.41184	11.12925	0	63.8
Quota present	3,015	0.793035	0.737392	0	2
Gender inequality	3,049	0.409229	0.194023	0.04	0.839
polity2	4,302	3.115295	6.641671	-10	10
GDP ($ billions)	4,538	244.5006	1101.103	0.008824	18036.6
GDP/cap ($ thousands)	4,535	11.18292	18.74049	0.06481	193.6481
Infant mortality	4,562	36.24851	33.65894	1.5	171.2
Ease start business	2,893	73.53564	18.3978	2.21	99.98
Credit registry	2,748	8.484425	17.89584	0	100

Acceptance of the value of gendered policy making is captured by considering whether the country has a gender quota for the national legislature. These quotas appear in a variety of forms. They may be voluntary quotas, legislated quotas or constitutional quotas. They may be quotas on the number of candidates or the number of actual seats. The Varieties of Democracy database has comprehensive information on each of these aspects of parliamentary gender quotas (Coppedge et al. 2017; Pemstein et al. 2017). We code the variable 1 if there is a legislated or constitutional quota on candidates, a 2 if a legislated or constitutional quota on the number of seats and a 0 otherwise. All other data comes from the World Bank.

Table 2: Regression Results

	(1)	(2)	(3)	(4)
	Microfinance Regulatory Environment			
% women MPs	-0.0333		-0.0446	-0.0270
	(-0.62)		(-0.84)	(-0.52)
Gender Quota		8.591^{*}	8.667^{*}	8.110^{*}
		(2.54)	(2.56)	(2.34)
Gender Inequality	16.73	21.55	18.19	34.72
	(0.77)	(0.97)	(0.81)	(1.52)
Democracy	1.043^{**}	1.563^{***}	1.525^{***}	1.466^{***}
	(2.70)	(3.68)	(3.58)	(3.43)

	(1)	(2)	(3)	(4)
	Microfinance Regulatory Environment			
GDP	0.00225	0.00215	0.00206	0.00159
	(1.86)	(1.81)	(1.73)	(1.36)
GDP/cap	0.0784	0.00989	0.0677	-0.172
	(0.24)	(0.03)	(0.21)	(-0.50)
Infant mortality	0.546***	0.456***	0.465***	0.436***
	(4.34)	(3.65)	(3.68)	(3.48)
Ease start business				-0.0248
				(-0.32)
Credit registry				0.242***
				(3.79)
Constant	21.53*	10.50	12.55	6.649
	(2.27)	(0.97)	(1.15)	(0.50)
N	261	232	230	230

t statistics in parentheses

$^{*}p < 0.05$, $^{**}p < 0.01$, $^{***}p < 0.001$

I use cross-sectional time-series regressions to test for a relationship between the ratio of women in parliament and the regulatory environment for microfinance. After controlling for basic socio-economic factors that might affect the need for microfinance, the ratio of women in parliament has no statistical effect on the regulatory environment for microfinance. There is no statistical support for hypothesis 1. While there is evidence from other research that women often taken on issues and champion causes that promote the welfare of women (Celis 2006; Williamson and Carnes 2013), they do not seem to push for better microfinance regulation. This is despite the general consensus that better regulation leads to more growth in the industry (Hartarska and Nadolnyak 2007; Arun 2005). On the other hand, some scholars have suggested the increased regulation, because it increases costs, might also reduce outreach to the poorest of the poor and lead to mission drift (Demirguc-Kunt et al. 2011; Espallier, Bert D; Guérin, Isabelle; Mersland 2011).

Hypothesis 2 is supported by the data though. Through a variety of configurations, having either a legislated or constitutional gender quota has a statistically significant impact on improved microfinance regulation. That is, countries that seem to be more gender aware, by having gender quotas in government, seem to be more likely to craft microfinance friendly regulations.

CONCLUSIONS

Considerable evidence has come to light over the last decade suggesting that microfinance might not be a very effective solution for addressing poverty (Roodman 2012; Warby 2015). However, microfinance is not an obscure poverty intervention. Quite the opposite. The microfinance revolution received a great deal of attention on global stages throughout the first decade of the 21st century. Between the United Nations' *Year of Microfinance*, Muhammad Yunus winning the Nobel Peace Prize, and countless books, articles and reports about microfinance, it was a movement that gained real traction. The industry also spread globally during the same time frame, reaching into nearly every country in the world. Policymakers around the world were quite aware of the industry.

Microfinance often has a clear gendered component. Muhammad Yunus typically framed his accounts of microfinance as lending to women. The literature is full of references to female empowerment and gendered lending programs (Mcintosh et al. 2013; Hermes, Lensink, and Meesters 2008; Imai, Arun, and Annim 2010; Khan 2009; Karlan and Goldberg 2011; Banerjee et al. 2013), and, although microfinance has not proven to substantially increase income, it has boosted freedom in women's spendings and earnings (Loiseau and Walsh 2015). There is also considerable evidence, as noted previously, that female MPs tend to champion causes that might be thought of as women's issues. However, they do not seem to fight for microfinance, at least not through fostering a friendly regulatory environment.

Perhaps microfinance regulation is not perceived as a women's issue by female MPs. It is often rather technical and does not capture the same passion as fighting for property rights, access to family planning, or other more immediate challenges that poor women often face. Indeed, a study of female legislators in Great Britain, shows that women in government tend to advocate for policies of female autonomy rather than economic policies, even when they might benefit women (Lovenduski and Norris 2003).

It is also possible that, while academics see the gendered aspects of microfinance, it also serves a lot of men, and many of the not-so-poor socio-economic classes. The evidence here does not suggest that female MPs do not champion women's issues, only that microfinance is not championed as a women's issue.

Interestingly, however, gender aware countries, that is countries that have gender quotas for their governments, are more proactive in fostering microfinance-friendly regulatory environments. Passing amiable regulations is not the same as putting significant government resources into increasing the outreach and impact of microfinance. However, it is not insignificant either. Better regulation has a positive effect on microfinance's impact for poor borrowers (Hartarska and Nadolnyak 2007; Arun 2005). It is important to acknowledge that development practitioners in a few countries have noticed that an exaggerated focus on microfinance tends to lead to a neglect of the employment and livelihood needs of the poor (Kumar 2013). Indeed, as mentioned above, microfinance has its challenges and it is far from being a cure-all solution. The idea of microfinance as a tool for empowerment must be accompanied by the understanding that empowerment is not limited to the access to tools but must involve a shift in power relations (Cornwall 2016; Duflo 2011; Malhotra, Schuler, and Boender 2002). The tool loses its effectiveness when it is not paired with a societal shift towards equality. This might explain why female MPs might prioritize policies of welfare and societal change to lift women out of an environment of gender inequality before focusing on empowering them financially. Although financial empowerment does contribute to this process, it might be ineffective on its own and can be more successful in combination with a system of gender equality already in place (Kabeer and Natali 2013). Without an ideology of social equality, control over financial resources is not enough. This might also explain why countries with gender quotas in place, and therefore a focus on gender equality, might champion microfinance friendly policies as a gendered issue.

Future research should look at why female MPs do not take on microfinance regulation. Is it a matter of perceptions? Or an issue of technocratic details? Or simply priorities? We should also continue to think about the much larger question, why do some governments seem to be more interested in helping the poor than other governments? In a world that is still struggling to address global poverty in its various forms, this is an important question that needs carefully considered answers.

Chapter 4

Corporate Social Responsibility–An Institutional Perspective

Satyanarayana Vadlakonda

ABSTRACT

Corporate social responsibility (CSR) refers to business practices involving initiatives that benefit society. This article maps out the crucial role and responsibilities for business in fighting poverty and acting responsibly in developing societies. What is clear from this article is that CSR in developing societies is a rich and fascinating area of enquiry, which is becoming ever more important in CSR theory and practice. And since it is profoundly under-researched, this enquiry also represents a tremendous opportunity for improving the knowledge and understanding about CSR. In the areas of Millennium Development Goals - Eradicate extreme poverty and hunger, Achieve universal primary education, promote gender equality and empower women, reduce child mortality, improve maternal health, combat HIV/AIDS, Malaria & other fatal diseases, ensure environmental sustainability, develop a global partnership, I offer an institutional theory of corporate social responsibility consisting of a series of propositions specifying the conditions under which corporations are likely to behave in socially responsible ways. I argue that the relationship between basic economic conditions and corporate behavior is mediated by several institutional conditions: public and private regulation, the presence of nongovernmental and other independent organizations that monitor corporate behavior, institutionalized norms regarding appropriate corporate behavior, associative behavior among corporations themselves, and organized dialogues among corporations and their stakeholders.

Keywords: Institution, Millennium Development Goal, Corporate Social Responsibility, Eradicate extreme poverty, Improve Maternal Health

INTRODUCTION

India is a country of many contradictions. In recent times it appeared to be one of the largest economic growths in the world, and it has come out an important one across the world as far as social growth is concerned. In spite of having great history in the world, the country has been suffering from social problems particularly poverty where a large number of people of the country are unable to access food even twice a day. Therefore, it has resulted in children malnourished vehemently. This is because of the social unrest in the country. When the development of the country has become crippled, the philanthropists emerged to extend their support in view of helping the needy people in the name Corporate Social Responsibility (CSR) in the year 2013. The primordial objective of this scheme is to eradicate poverty and banish the malnourishment among the poor children in the country over period of time. Ever since the CSR was established, there was an enthusiastic response from many corporate companies to draw the attention of the governments and to earn an international reputation.

Numerous NGO's have come forward to help the hapless citizens who are under the below poverty line. Their contribution to the society is growing rapidly day by day. Subsequently, the social expectation was built up among the haves and have-nots towards the social group. Over the passage of time, other companies have emulated this social cause of special assistance for the needy and poor in various parts of the country. Companies thus played an instrumental role in prevention of social problems to some extent. In response to CSR, the governments and regulators have set up the scheme under the title of National Voluntary Guidelines for Environmental and Economic Responsibilities of Business or the NVGs (accompanied by the Business Responsibility Reports mandated by the SEBI for the top 100 companies) and the CSR clause within the Companies Act, 2013, these are two such steps taken.

According to Indian Institute of Corporate Affairs, a minimum of 6,000 Indian companies would be required to undertake CSR projects in order to comply with the provisions of the Companies Act, 2013. Later, a good number of companies undertook initiative for the promotion of this scheme. Further, some estimates indicate that CSR commitments from companies could amount to as much as 20,000 crore INR. This combination of regulatory as well as societal pressure has meant that companies have to pursue their CSR activities more professionally.

METHODOLOGY & OBJECTIVES

The main objective of this paper is to examine the corporate social responsibility in India as well as in the case study of NTPC Ramagundam. The present empirical study focuses on NTPC CSR Ramagundam. The secondary data has been collected from CSR CD department- NTPC Ramagundam and census in 2011. The various reports have been analyzed in a comprehensive manner in the study.

SOCIO-ECONOMIC BACKGROUND OF THE STUDY AREA

NTPC, a Government of India undertaking has setup a Thermal Power Station 1978 with a capacity of 2600MW at Ramagundam, Karimnagar District of the then Andhra Pradesh.(Now it is in Telangana Sate) Ramagundam is situated at a distance of about 75 Kms from Karimnagar. The powerhouse at Ramagundam is the biggest thermal power producing station in the state and supplies power to Hyderabad, Karimnagar, Warangal and Mancherial Cement Factory. A Navaratna Public sector undertaking completed 36 glorious years in service to the Nation (CSR- CD NTPC Ramagundam, 2016).

The study explores the NTPC-CSR areas like Ramagundam and Kamanpur selected for the research which are very nearby and adopted by NTPC, the total population of these two Mandals 3,27,167 (as per 2011 census). The sample for research has been taken from 50% of the villages adopted by the NTPC for CSR – CD programmes randomly.

CSR : Community Development Programmes of NTPC

- Promotion of education
- Promotion of Health and creating awareness on health seeking behaviour
- Providing infrastructure for sustainable development
- Enhancing vocational skills and environment sustainability etc.,
- Women Empowerment & Gender equity
- Other Welfare Activities

CSR Activities in Ramagundam

The company has introduced CSR policy in the year 2004 and the same policy has been revised after 10 years that is in 2014. Mostly the policy of CSR is aimed to benefit the neighboring villages. The company so far has identified some welfare activities like safe drinking water, primary health, education, including vocational trainings and roads and environmental issues in the surroundings.

Table 1: CSR Budget of NTPC Ramagundam

Sl. No.	*Year*	*Amount (in laks)*	*Sl. No.*	*Year*	*Amount (in laks)*
1	2004-05	31.6	7	2010-11	179.99
2	2005-06	18.61	8	2011-12	319.74
3	2006-07	54.14	9	2012-13	694.31
4	2007-08	44.12	10	2013-14	460.21
5	2008-09	37.23	11	2014-15	971.45
6	2009-10	101.49	12	2015-16	1845.57
Total					4758.02

Source: CSR – CD Department NTPC.

As per Table 1 in the year of 2004-2005 financial year the company has spent Rs 31 lakhs 16 thousand only. Year by year the amount has been increased and in 2014-2016 financial year the total amount spent on this account is 18 crores 45 lakhs 57 thousand. The company had spent 47 crores 58 lakhs in last twelve years. The entire budget is dedicated to community development projects in the targeted villages. In the beginning the company had concentrated on rehabilitation and resettlement activities later the entire endeavour is to improve the quality of life of the people in surrounding areas of NTPC plant. The community development projects aimed at the basic amenities like education, infrastructure, sports and cultural activities, livestock and other awareness programmes.

Education

Education is an effective tool for bringing social change. NTPC Ramagundam has adopted a multi pronged approach to promote quality education in surroundings villages. The basic activities under taken by the NTPC are primary school education, construction of compound walls for government schools, distribution of note books, merit scholarships, uniforms and school bags, carrier guidance to students, girls toilets, furniture, ceiling fans, donating books to library etc.

Table 2: Education Activities

Sl. No.	*Education activities*	*Total no. of villages covered*	*Total no. of beneficiaries covered*
1	Distribution of books	235	29,132
2	SSC Merit scholarships	114	500
3	Uniforms and School bags	72	5,527
4	Adult education in the nearby village	9	278
5	Career Guidance to Students	30	5,176
6	Construction of Compound wall gate etc.	9	760
7	Construction of Cycle shed near FCI, Elakapalli	1	108
8	Dues desk benches	26	3,144
9	Repairs in Govternment schools at Kannala	-	299
10	Solar Lanterns	32	546
11	Girls Toilets, Repairs of Schools	24	9,560
12	Furniture to Schools	51	1,458
13	Construction of Girls Toilets	2	470
14	Ceiling fans (Anganwadi)	839	8,390
15	Support to inclusive Education	2	10
16	Support Library Activities to Govt. Schools	25	16,765
17	Renovation of ZPHS School	1	146
18	Construction of Amphi Theater Godavarikani	3	850
19	Water filters to Government Primary School Ramagundam	50	4,413
20	Expenditure made on village students in NTPC	12	2,867
	Total		90,159

Source: CSR – CD Department NTPC, Ramagundam

Health

The hygienic conditions in rural areas mostly cannot support the health of the people. Due to the poor hygiene conditions people in rural areas suffer from ailments and diseases. NTPC under CSR-CD has been trying to create health and also providing awareness, necessary medical help. Several health camps were conducted such as eye care, dental care and general medical camps etc. The only nearby available 100-beded hospital is situated at Godavarikhani apart from Singareni collieries hospital and NTPC hospital and other State Government PHC's. There is an NGO Sai Seva Samithi catering to the medical needs of poor people free of cost, which receives support from NTPC. Medical shops are available at Urban Centers only.

Table 3: Health Activities Conducted by NTPC

Sl. No.	*Health Activities*	*Total no. of villages covered*	*Total no. of beneficiaries covered*
1	Construction of Health Centers	5	1147
2	Medical Camps for Public, School, Eye Camps	63	28422
3	Eye camp Glasses, Family Planning Operations	22	2508
4	Medical Camps for Public, School	25	5663
5	Mega Eye camp at Kundanapalli	1	150
6	Mega Health Care on respiratory diseases PK Ramaiah Colony	1	260
7	Distribution of MCR footwear for Leprosy Persons	1	300
8	Special Surgical Camp for Villagers, Gynic	5	15
9	Installation of solar plant to Govt. Hospital	1	7200
10	Providing SDP Platelet Machine to Red Cross	1	600
11	Community Toilets at Medipalli Center	1	18000
12	Support to Area Hospital, Godavarikani	37	68963
13	Pulse Polio	35	3985
14	DOT Centers	38	6874
15	Cleaning drains	3	2000

16	Development of drain system	1	337
17	Construction of Individual toilets	2	55
18	Open drains in urban areas	2	6000
19	Awareness of mother and child health programme	13	272
20	Cleaning of natural nalla from plant to Newpali	3	2500
21	Renovation of Malyapalli pond	1	2500
	Total		159700

Source: CSR – CD Department NTPC, Ramagundam

Drinking Water Activities

Under CSR-CD drinking water projects the NTPC has established reverse osmosis plants (RO's) in several villages apart from supplying drinking water through tankers. In few villages it also supported to drill bore wells under the safe drinking water projects around 79,850 people were benefited.

Table 4: Drinking Water Activities in CSR Scheme

Sl. No.	*Drinking Water*	*Total no. of villages covered*	*Total no. of beneficiaries covered*
1	Drinking water supply through tankers	7	15000
2	Pipe line extension at Kaziplli SC Colony	1	390
3	Pipe line extension at Malyalpalli	1	353
4	Drinking water scheme for Kundanpalli	1	4000
5	Supply of drinking water, Bore well etc.	10	2000
6	Providing Reverse Osmosis Plant	5	12201
7	Water supply scheme Kundanpalli	2	10000
8	Yelampalli project & RO Plants	5	6310
9	Ro plants in Elkalapally, Annapurna colony	2	2300
10	Solar Power in Community Centers	5	3500
11	RO Plants at Milliapalli	2	1196
12	Bore wells 13/Supply of drinking water	12	9750

13	RO Plant at P.K. Ramaiah Colony, Yellampalli, Murmur	4	600
14	Bore wells Elkalapalli, Laxmipuram, Kazipalli	5	1250
15	Support to drinking water system at Lingapur	1	1000
16	Motorized bore wells & hand pumps	10	8000
17	RO Plant at Julapalli	1	1000
18	RO Plant at Vemnur and Potyala	1	1000
	Total		79850

Source: CSR – CD Department NTPC, Ramagundam

Activates for Differentially Able Persons

NTPC Ramagundam has organized many activities to differentiate able persons such as providing free education, distribution of weal chairs and tricycles. The people benefited from the programme are around 1424.

Table 5: Support to Differentiate Able Persons

Sl. No	*Support to DAP*	*Total no. of villages*	*Total no. of beneficiaries*
1	Inclusive education for Dap students	3	12
2	Issue of Ash bricks to Manochaitanya	1	100
3	Inclusive education for DAP students in townships	3	21
4	Support to DAP Schools at Karimnagar	1	100
5	Support to DAP children for education	4	11
6	Support to Government ITI Karimnagar	1	17
7	Tri Cycles & Wheel-chairs	5	372
8	Inclusive education	7	36
9	Paper cup making machine Manochautanya	1	100
10	Purchase of mini bus for Manivikas, Karimnagar	1	80
11	Distribution of tricycles 22 & wheel chairs 17	10	39

12	DAP inclusive education	1	6
13	Support to Manovikas Students at Karimnagar	1	100
14	Construction of addl. Rooms at Manochaitanaya	1	100
	Total		1424

Source: CSR – CD Department NTPC, Ramagundam

Live Stock Activities

Several live stock health camps were conducted in association with State Animal Husbandry Department nearby villages to promote their good health.

Table 6: Live Stock Activities

Sl. No.	*Live Stock*	*Total no. of villages covered*	*Total no. of beneficiaries covered*
1	Live Stock health camps	10	653
2	Live Stock health camps	10	741
3	Live Stock health camps	10	518
4	Agricultural appliances to Gramapanchayats	13	566
5	Veterinary Centers at Rampur & Ramagundam	22	6120
6	Live Stock Camps	10	10558
7	Live Stock Camps, Insurance	10	675
8	Agriculture Appliances training	15	240
9	Animal health & agricultural appliance	10	547
	Total		13,188

Source: CSR – CD Department NTPC, Ramagundam

CONCLUSION

As far as education is concerned, NTPC extended the rudimentary facilities particularly providing infrastructure, books, study material and play material and laboratory equipment in a satisfying manner. Health issues have been sustained greatly by providing and strengthening institutional deliveries to decrease infant & maternal mortality and by also creating awareness on routine immunization. Conducting awareness programs on personal hygiene, ecological balance, seasonal and chronic diseases like HIV, Malaria, Dengue, Chickun Guinean, polio etc. and other facilities have been provided to the inhabitants of the chosen places appropriately.

Collaboration with Singareni Collieries Company Limited is very essential in implementing CSR – Community Development activities. Both the companies are implementing their CSR activities independently. Greater impact can be expected if these two big players can do the same collectively. But, the promises which were made during its inception have not been enter fulfilled. To fulfill the CSR – CD in a systematic way NTPC has to engage the professional social workers for its CSR – CD instead of MBA and other graduate only.

There is a need to create awareness about the bad effects of alcoholism and youth must be given some skill oriented vocations, as most of the youths are unemployed in these villages, though they were qualified in formal education like intermediate and graduation. There is a need to sensitize people about self-employment and the importance of work and entrepreneurship. Though it (NTPC) is achieved to some extent, it's CSR – Community Development Program, but it is unable to fulfill the needs and requirement of the villages as desired by villagers. At the time of its establishment, people had lot of hopes they had sacrificed their cultivable lands and villages for this construction. Majority of the villagers have been displaced during its establishment.

NTPC has to work for its promises which were made during its establishment and see that the basic needs of these villages must be met more extensively and beneficially.

REFERENCES

1. Corporate Social Responsibility: Background and Perspective, John Samuel and Anil Saari.
2. Hand Book on Corporate Social Responsibility in India.
3. http://ec.europa.eu/enterprise/policies/sustainable-business/corporate-social-responsibility/index_ en.htm
4. http://www.wbcsd.org/work-program/business-role/previous-work/corporate-social-responsibility. aspx http://www.unido.org/what-we-do/trade/csr/what-is-csr.html#pp1[g1]/0/
5. http://www.soulace.htm (accessed on 31.03.2017 at 8.46pm). 7 From a study conducted by Ernst & Young and PHD chamber, published in Global CSR Summit-2013

Chapter 5

Sustainable Human Development Poverty Reduction and the Elimination and Inequality

Suman Maurya

ABSTRACT

Good Governance and Human Rights are mutually rein- forcing–both being based on core principles of participation, equality and dignity. Human Rights principles provide a set of values to guide the work of governments and other socio-political spheres. Good governance and respect for Human-Right are essential for sustainable human development, poverty reduction and the elimination of inequality. Good governance is associated with efficient and effective administration in democratic framework considered as citizen-friendly, citizen-caring and responsive administration that projects and promulgates the universality of human rights such as civil, cultural, economic, political and social rights. Human - Rights identifies five key attributes of good governance that are transparency, responsibility, accountability, participation, and responsiveness.

It is now being observed that protection of human rights at every level is possible only when good governance prevails. Good Governance can promote different human rights from right to life to environmental rights as well as socio-economic-political rights. It is therefore demanded to declare right to good governance as most basic human right. The paper attempts to analyze the conceptual background of these two concepts.

Keywords: Good Governance, Human Rights, Sustainable development, Accountability, Democratic-framework.

GOOD GOVERNANCE: EVOLUTION AND ATTRIBUTES

Good Governance and Human Rights are mutually reinforcing. Good governance is associated with efficient and effective administration in a democratic framework. It is considered as citizen friendly, citizen caring and responsive administration. Good governance emerged as a powerful idea when multilateral and bilateral agencies like the World Bank, UNDP, OECD, ADB, etc realized that project's success substantially depended on conditions of governance in the aid receiving countries. As the UN Secretary General Kofi Annan mentioned that "Good Governance is perhaps the single most important factor in eradicating poverty and promoting development".

In general, good governance is perceived as a normative principle of administrative law, which obliges the state to perform its functions in a manner that promotes the values of efficiency, non-corruptibility and responsiveness to civil society. United Nation Development Programme while defining concept of good governance placed greater emphasis on sustainable human development. The UNDP emphasizes the human development and elimination of poverty and public administration.

Govern or Governance means rule or control with authority, conduct the policy and affairs of government, influence or determine a course of action, be the predominating influence, be a standard or principle for; serve to decide, Good Governance is synonym of Good Management for India. According to the World Bank Governance is "the manner in which power is exercised in the management of a country's economic and social resources for development". The use of the term "Good Governance" was initially articulated in a 1989 World Bank Publication. The concept of Good Governance in International law has also been well received under the rubric of human rights. Increasingly, in international development literature, the concept of good governance parallels a normative (rule-setting) formulation, which reflects the degree to which it promulgates the universe of civil, cultural, economic, political and social rights. The United Nations High Commissioner for Human Rights (UHCHR) identifies five attributes of good governance as follows: Transparency; Responsibility; Accountability; Participation; and Responsiveness. The concept of governance is as old as human civilization. 'Governance' simply means the process of decisions is implemented. On the relationship between human rights and governance, the bank conceptualized governance to indicate the manner in which rights associated with power and authority are exercised for development in the management of a country's economic and social resources. According to the International Monetary Fund, Good Governance is important for countries at all stages of development it emphasizes on its importance in fields such as promoting public sector transparency and accountability. According to the UN Economic and Social Commission for Asia and the Pacific, good governance has eight major characteristics. It is participatory, consensus oriented, accountable, transparent, responsive, effective and efficient, equitable and inclusive and follows the rule of law. In the first instance the term Good Governance was derived from international arena as a part of initiatives of the World Bank to ensure that the development assistance is used effectively. But it has broadened its dimension, including the political and social aspects especially with the emergence of the concept of sustainable development by the mode of implementing Human Rights. (Mihar 2017; 72-80) It assures that corruption is minimized, the view of minorities are taken in account and that the voice of the most vulnerable in society are heard in decision making. It is also respective to the present and future needs of society.

Human Rights and Good Governance : Interconnectedness

Human Rights are those inalienable and essential rights which are universally possessed by each and every human being by virtue of being human. These are set out in the Universal Declaration of Human Rights of 1948 and codified and further spelled out in series of international conventions with the objective of ensuring human dignity and democratic values. Human Rights principles provide a set of values to guide the work of governments and other political and social actors. It establishes the equal dignity and worth of every person. It confirms that the state has a core duty to promote standards of life that enable us to exercise our dignity and equality, in larger freedom. Human rights are relevant for all of us and every day. Human rights include our rights to freedom from fear and from want, freedom to speak up, rights to health and education and to enjoy the benefits of measures to advance economic and social justice. Our shared humanity is rooted in these universal values. (Neier, 2012; 60-71) Governance and human rights are inter-connected the fulfillment of protection of human rights is benign responsibility of government so that these are blended together. Respect for human rights is core principle for sustainable development, global peace and inclusive growth. Protection of human rights should be done locally as well as globally by the means of good governance.

The Karnataka Lokayukta Hon'ble Mr. justice N. Santosh Hegde in the first CIPS Foundation Day Lecture said that good governance is a right of citizen and democracy. Such governance includes factors such as transparency and accountability. It also includes values such as justice and equity. It must ensure that citizens, especially the poorest, basic needs are met and they live a life with dignity. Good governance implies an administration that is sensitive and responsive to the needs of the people and is effective in coping with emerging challenges in society by framing and implementing appropriate laws and measures. It includes strict rules of accountability. Common good has led to evolution of good governance as opposed to mere governance, as the umbrella concept, encompassing within a system and principle of "Welfare State" that is able to unequivocally discover the basic values of the society where standards concern economic, political and socio-cultural issues including those involving human rights and follows the same through an accountable and upright administration.

National Human Rights Commission: Contribution to Good Governance

The statutory public body National Human Rights Commission (NHRC) of India constituted in 1993 is responsible for the protection and promotion of human rights granted by the constitution. In consonance with Paris principles the NHRC functions to ensure accountability and strengthening of human rights in the country.

Apart from looking into complaints of the Human rights violations, the commission's functions include reviewing safeguards provided under the Constitution of Law, and also make recommendations for effective implementation of international conventions/covenants. The NHRC is entrusted with the responsibility of undertaking research and to organize seminars and discussion programmes on human rights issues, spread awareness about human rights and encourage efforts of nongovernmental organizations towards promotion of human rights. It has also been organizing camp sittings for disposal of pending cases and open hearings of complaints of atrocities against Scheduled Castes and Scheduled Tribes in different parts of the country for the redressel of their problems. If has also been visiting, state wise different districts to make an assessment of enforcement of various majors benefits and entitlements related to human rights and implementation of welfare schemes so as to come out with recommendations for the governments, committed to ensuring good governance. It has made a number of interventions on key issues of human rights which includes, among others, issues of bonded and child labour, prison reforms, right to health, right to food, mental health care, rights of persons with disability, silicosis, illegal clinical drug trials, pesticides in food items, pricing of drugs, corporate-social responsibility, manual scavenging and sanitation, human rights of women. (Taylor 2004; 32-34)

In order to spread awareness about human rights and its activities, the Commission, apart from publication of monthly newsletter in Hindi and English, has published more than eighty books and journals. Besides the Central and state governments, several organizations, NGOs of human rights defenders and media have actively supported and supplemented the work of the NHRC towards promotion and protections of human rights.

Human rights and development aims converge in many instances and are often beneficial only to the government and not the people, although there can be conflict between their different approaches. Article 21 of UNHRC is the right to participate in government that recognizes the importance of participatory government and Article 26 provides Right to Education which is first step for good governance. Article 28 importantly signifies– everyone is entitled to a social and international order in which the rights and freedoms set forth in this declaration can be fully realized.

The UN charter allows for a limit to state sovereignty where human rights are threatened. Two critiques of that are that it did not make political rights dependent on multi party democracy and there is a lack of protection for ethnic minorities. Protecting individual rights do not necessarily protect group rights.

The nexus between grave human rights violations and international security is significant as atrocities within a sovereign state are of concern to International law, when they upset neighboring states in a manner disturbing world peace. Taken together the United Nations Charter and Universal Declaration of Human Rights provide a legal mechanism which may challenge the sovereign rights of states to oppress people within their own jurisdiction. (Oliver 2008; 41–43)

The Right To Development: Implications

The major human rights principles guiding the programme are regarded as universality and inalienability, indivisibility, interdependence and interrelatedness, non-discrimination and equality, participation and inclusion, accountability and the rule of law. The Right to development is associated both governance as well as Human Rights. The Right to development is regarded as an alienable human right which all peoples are entitled to participate in, contribute to and enjoy economic, social, cultural, and political development. The right to development consists of people centric development, participation, equity, non-discrimination. The right generation right viewed as a group right such that it is owed to communities as opposed to an individual right applying to individuals. Most developing states voice concerns about the negative impacts of aspects of international trade, to technology and crushing debt burden and hope to create binding obligations to facilitate development as a way of improving governance and rule of law.

In September 2000,world leaders made in the Millennium Declaration UN the Resolution to formulate governance that includes peace, security, human rights, the environment and development targets which were later configured into Millennium Development Goals (MDGs). The Millennium Declaration considers six fundamental values necessary for international relations such as Freedom to raise children in dignity, freedom from hunger and from the fear of violence, oppression and injustice, including democratic and participatory governance based on the will of the people, Equality–no individual or nation must be denied the opportunity to benefit from development. Solidarity with principal of equity and social justice, Tolerance that is cherished as a precious asset of humanity, Sustainable development and unsustainable patterns of production and consumption must be changed in the interest of the future welfare of our descendants. Shared Responsibility is responsibility for managing worldwide economic and social development. Human rights have played limited role in influencing MDG planning, though there are strong similarities between them and the content of MDGs which resemble many economic and social rights. MDGS provide benchmarks for economic and social rights, while human rights strategies offer enhanced legitimacy, equity and sustainability to the MDG policies. The Millennium Declaration substantially refers to human rights and governments have committed themselves to respecting recognized human rights and fundamental freedoms, including the right to development. Economic, social, and cultural rights, the rights of women, migrant, minorities, and participation are all emphasized.

The economic and social rights, such as the right to health care or education, may be understood not as legal instruments for individuals, but as duties for governments and international agencies such that everyone bears some responsibility for the fulfillment. Economists accept that the realization of high standards of health and education are conducive to economic growth. The Human Rights approach regards transparency and empowerment as ends in themselves, while an economic approach sees them as instrumental to a welfare outcome.

Significance of Strategies and Policies

The government must amend laws to include a timeframe to carry out work in transparent manner that ensures accountability. Penal clauses must also be introduced so that action could be taken against the official. Provision must be made for compensating the wronged citizen making good governance a right, widening the scope of the definition of corruption and creating the Lokpal institution should be accountable to the Lok Sabha and the general public. (Kumar 2011; 23–25) According to the UN High Commissioner for Human Rights, the true test of "good" governance is the degree to which it delivers on the promise of human rights :civil, cultural, economic, political, and social rights. In spite of that the key question still remains are the institutions of governance effectively guaranteeing the right to health, adequate housing, sufficient food, quality education, fair justice and personal security?

The concept of good governance has been clarified by the work of the commission on Human rights. Its resolution 2000/64 expressly linked good governance to an enabling environment conducive to the enjoyment of human rights and "prompting growth and sustainable human development".

By linking good governance to sustainable human development, emphasizing principles such as accountability, participation and the enjoyment of human rights, the resolution stands as an impact endorsement of the rights based approach to development. In its resolution 2001/72, the Commission on human rights reaffirmed its earlier resolution by consensus. The following section builds on the development process.

The UNDP Human Rights Development Report, 2002 elaborates 'Democratic Governance' as governance that would promote human development. Like the concept of good governance ;democratic governance seeks efficient institutions, and a predictable economic and political environment necessary for economic growth and effective functioning of public services. But the concept of democratic governance concerns political freedom and human rights, and removal of discrimination as central objectives. A reform agenda would aim at building institutions and rules that are not just efficient but also fair, and that are developed through a democratic process in which all people have a real political voice. Democratic governance thus incorporates into the notion of good governance for development, democratic processes and institutions, and a concern for the securing of political and civil rights and freedoms as human rights. (Donnelly 2003; 22-31)

All these expressions convey theories pregnant with time tested concepts. The participation in order to be effective, needs to be informed & organized and, therefore, depends upon the availability to the subject "freedom of association & expression" on one hand and existence of "an organized civil society" on the other. This necessarily is apointer to representative democracy. The attribute of "rule of law" inheres as prerequisite "fair legal frameworks" that are enforced impartially and particularly "full protection of human rights", especially of the vulnerable sections of society. The factor of "transparency" requires that information is freely available and the decisions are taken or enforced in a manner that adheres to the rules and regulations. The attribute of 'responsiveness' necessitates that all public institutions and their processes strive "to serve all stake holders within a time frame". Democracy, liberty and the rule of law together represent the good governance that is universally accepted now as the index of a civil society. Democracy signifies a government of, by and for the people. (Reif 2013; 125-128)

The protection of individual liberties follows the notion of democracy as a natural corollary. This entails the espousal of a methodical configuration of laws by which society might be regulated and different conflicting interests can be harmonized to the full extent. This is why "the rule of law" is indispensable. It envisages the pre-eminence of law as opposed to anarchy or capricious dictates. It involves equal accountability of all before the law irrespective of high or low status.

CONCLUSION

Democracy has been evolved through centuries of experience amongst the people, who care for human person,dignity and rights as the best and most acceptable form of good governance. It is a concept that highlights on this idea that all citizens have a right to participate in the decision-making processes that leads to adoption of policies are applicable to the societies.

In India basic human rights are incorporated as guaranteed fundamental rights, elaborated in every possible manner, in part III of the constitution. These fundamental rights go much beyond the American Bill of Rights. They did draw upon the Universal Declaration of Human Rights issued by the United Nations in 1948 but went ahead of them by incorporating alongside, in Part IV of the constitution, certain 'Directive Principles of State Policy' which are principles that would be fundamental for "good governance" of this country. The Directive Principles have been used as fundamental principles of governance tempered by the Fundamental rights and Directive principles that can ensure dignified growth and development of each individual as a useful human being. (Hunt 2007; 176-178)

Human rights principles provide a set of values to guide the work of governments and other political and social actors. They also provide accountability and human rights principles inform the content of good governance efforts. However, without good governance,human rights cannot be respected and protected in a sustainable manner.

From a human rights perspective, the concept of good governance can be linked to principles and rights set out in the main international human rights instruments. When led by human rights values, good governance reforms of democratic institutions create avenues for the public to participate in policymaking either through formal institutions or informal institutions. They also establish mechanisms for the inclusion of multiple social groups in decision making processes, especially locally. Finally, they may encourage civil society and local communities to formulate and express their positions on issues of importance. (Agere 2000; 11-18) Good governance reforms advance human rights when they improve the state's capacity to fulfill its responsibility to provide public goods which are essential for the protection of a number of human rights, such as the right to education, health and food. Reform initiatives may include mechanisms of accountability and transparency, culturally sensitive policy tools to ensure that services are accessible and acceptable to all, and paths for public participation in decision making. Human rights sensitive good governance initiatives reform legislation and assist institutions ranging from penal systems to courts and parliaments to better implementation of rule of law and legislation. Good governance initiatives may include advocacy for legal reform, public awareness-raising on the national and international legal framework and capacity-building or reform of institutions. Good governance efforts rely on principles such as accountability, transparency, and participation to shape anti-corruption measures. Good governance will ensure equitable sustainable development marked by active citizens participation, absence of injustice and corruption and in this way contribute towards protection of human rights. (Zhang 2016; 103-105)

By enforcement and implementive measures countries and societies have come closer to achieve good governance in its totality however, to ensure sustainable human development and inclusive growth, actions must be taken positively for formulating the aim of good governance.

REFERENCES

1. Anja Mihar, (2017), *Cyber Justice: Human Rights and Good Governance for the Internet*, Ist Edition, Springer publications, Berlin, Germany, pp. 72-80
2. Aryeh Neier,(2012), *The International human Rights Movement: A history*, Princeton university press, Princeton, New Jercy pp. 60-71
3. Barbara E. Taylor, Richard Chait,& William P. Ryan,(2004),*Governance as a Leadership: Reframing the work of non profit*, John wiley & sons, hoboben, Newjersey, pp. 32-34
4. Caroline Oliver,(2008), *Getting started with Policy Governance: Bringing Purpose, Integrity, and efficiency to your Board*, Jossey-bass, A wiley print publications, Sanfrancisco, pp. 41-43
5. C. Rajkumar Kumar, (2011),*Corruption and Human Rights in India: Comparative perspectives on transparency and Good Governance*, Ist Edition, Oxford University Press, New Delhi, pp. 23-25
6. Jack Donnelly, (2003), *Universal Human Rights: In Theory and Practice*, 2nd Edition, Cornell University Press, London, pp. 22-31
7. Linda C. Reif, (2013), *The Ombudsman, Good Governance and the International Human Rights System*,12th Edition, Springer publications,Harvard, pp.125-128
8. Lynn Hunt,(2007), *Inventing Human Rights: A history*, 2nd Edition, W.W. Norton & Company, Newyork, pp. 176-178
9. Sam Agere, (2000), *Promoting Good Governance: Principles, Practices, and Perspectives*, Ist Edition, commonwealth secretariat, London, pp. 11-18
10. Wei Zhang, (2016), *Human Rights and Good Governance*,Ist Edition, Martinus Nijhoff Publishers, Boston, pp. 103-105

Chapter 6

Women as Change Agents of Sustainable Development in South Asia

Snehil Kacker

ABSTRACT

Conflict, change, crisis affects men and women differently. It is an accepted fact that despite being equal to men, women are more vulnerable to exploitation and torture. The condition of women has worsened than before because of natural and man made disasters. Climate change and globalisation have brought about rampant changes in the lives of womenfolk. It is true that they are the victims of such changes but have also become the agencies of change,either by choice,by force or by reluctance. They have become agents of peace and development. Women have been the traditional guards of nature and have contributed in the most constructive manner towards its preservation. The paper discusses the impact of climate change on women in South Asia and also presents an alternative discourse on the importance of women in sustainable development.

Keywords: *Women, Sustainable Development, Agents, Climate Change.*

CLIMATE CHANGE

Climate forms the basis to exist on earth. The development of life forms and human civilization would not have been possible without a complex climate system. To most people, climate seems a simple process. However it is a complex system, involving many physical, chemical and biological interactions among air, water, ice, land surfaces, plants and animals.

It is a significant and lasting change in the statistical distribution of weather patterns over periods ranging from decades to millions of years. It may be a change in average weather conditions or the distribution of events around that average (e.g. more or fewer extreme weather events). This change may be limited to a specific region or may occur across the whole earth. Fluctuations, shorter than a few decades such as El Nino, cannot be called Climate Change.

Impact of Climate Change on South Asia

Climate Change has been one of the most challenging problems faced by mankind. It has been responsible for widespread impact on various aspects of life. All countries are affected by this problem in some form or the other; however there are some regions on the earth which are more affected than others. South Asia is one of the most vulnerable regions of the earth to the devastating impact of Climate Change.

Though South Asia occupies only 4.8 percent of the world's land mass, it is home to about 20 percent of the world's population. Population growth is directly related to deforestation, land degradation, food security and decline of air quality. South Asia is already a part of the global drivers of change like economic growth and globalization. This has resulted in large scale migration from rural to urban areas. Therefore, if on one hand South Asia boasts of Delhi, Mumbai, Kolkata, Karachi and Dhaka as the largest agglomerates in the world in 2011 (South Asia Environment outlook 2014, UNEP). On the other hand the same cities also harbor large slum areas. Dhaka is the fastest growing city in Asia yet it harbors 34% of the country's population. About half of the population of Delhi lives in slums and in case of Mumbai, the percentage of population living in slums can be 60%. (www.theguardian.com, 2012)

Statistics prove that the economic activity in South Asia will be adversely influenced by Climate Change. According to (Kumar and Parikh, 2001) even after accounting for farm level adaptation, a 2°C rise in mean temperature and a 7 % increase in mean precipitation will reduce net revenues by 8.4% in India. It is predicted that Pakistan will witness wheat yields decline by 6-9 % in sub humid, semi-arid, and arid areas with an increase of 1°C in temperature (Sultana and Ali, 2006). As for Sri Lanka,an increase in the frequency of droughts and extreme rainfall events could result in a decline in tea yield, which would be the greatest in regions below 600 metres'(Wijeratne, 1996) In Bangladesh,it is estimated that there will be net negative effect on the yields of rice, the staple food of the population. (Karim, Hussain and Ahmed,1996). (www.theworldbank.org)

Statistics prove that the population in South Asia has grown by 130 million during 2001 and 2011 and is supposed to grow to almost 250 million by 2030. Millions of South Asians live in informal urban settlements i.e. poor construction and under serviced plots. At the same time, the agglomeration economies of the region have two important achievements in the form of services and manufacturing which account for 80% of the region's GDP.

Therefore, it is important that South Asia does not lose the gains achieved after a long struggle, especially those achieved in the economic sphere.

IMPACT OF CLIMATE CHANGE ON SOUTH ASIA

Impact on Agriculture

Agriculture is the most affected sector on account of climate change due to its high dependence on climate. People involved in agriculture are generally poorer compared with urban residents. More than 60%of population is directly or indirectly relying on agriculture as a source of livelihood in this region. Climate change is already making a huge impact on the lives of the people particularly the poor. It is evident in a number of ways. Extreme warming trends and more frequent and intense weather events such as droughts, cyclones, floods and hailstorm have been observed across Asia.

Impact on Livelihood

The short term losses in the form of reduction in livelihoods and incomes due to impact of climate change on agriculture are well known. The short term costs of extreme climate events such as droughts, floods and cyclones can have major consequences for human development. In Asia, majority of population live on less than US$2 a day. Hence climate shocks can begin powerful **downward spirals** in human development. Climate shocks affect as they wipe out crops, reduce opportunities for employment, increase food prices and destroy property. However, the coping mechanism of rich and poor differ. Rich people can cope with shocks through private insurance or depending on their savings, the poor have no alternative but to reduce consumption nutrition, sell their productive assets on which their recovery depends. (Dev 2011; 1-11)

Impact on Pre- existing Conflicts

Climate change could accentuate pre-existing conflict patterns. By itself, climate change may not be sufficient to cause large scale security problems for South Asia. If we consider India, it is already exposed to potential internal and external conflicts. Climate change will heighten pre-existing conflict patterns.

In India,a major intra state conflict is naxalism. It is a reflection of the lopsided economic development in India. As resources become scarce and population increases, various conflicts will take place on resource allocation and distribution. Climate change will intensify dislocations. Naxalites are likely to attract more adherents as the stress of climate change mount.

Water Conflicts in South Asia

Climate Change also intensifies conflicts over water resources in South Asia. The implementation of agreements and treaties governing sharing of water get obstructed due to historical mistrust and aggrandizing control over scarce water resources. India-Bangladesh, India-Nepal, India-Pakistan already suffer from political turmoil due to water conflicts. The Indus water treaty of 1960 aims at sharing of water of Indus river between India and Pakistan. It provides three eastern rivers to be used by India and three western rivers to be used by Pakistan except for salient situations. (Rai and Patnaik,2011).Despite the treaty, its interpretation creates tension between the two countries. Pakistan has been raising enormous concern about the construction of dams on Indus by India. Similarly, India- Nepal have been involved in conflicts over water. These conflicts deepen with recurrent floods in the river Koshi. The floods in 2008, displaced 3 million people in India and 50,000 people in Nepal. Both countries blamed each other for poor disaster management (Malhotra, 2010).

If water conflicts get intensified by Climate Change, carbon emissions intensify climate change. In the wake of rampant industrialization and economic growth, countries in South Asia are incessantly increasing their carbon emissions. India, on account of its carbon emissions has been strictly monitored and pressurized by the international actors.

The table below shows carbon dioxide emissions of 20 countries in 2016. The data compiled by the International Energy Agency, which estimates carbon dioxide (CO_2) emissions from the combustion of coal, natural gas, oil, and other fuels, including industrial waste and non-renewable municipal waste.

Table 1: Total CO_2 Emission

Rank	*Country*	*Total CO_2 Emission*
1	China	9056.8 MT
2	United States	4833.1 MT
3	India	2076.8 MT
4	Russian Federation	1438.6 MT
5	Japan	1147.1 MT

Rank	*Country*	*Total CO_2 Emission*
6	Germany	731.6 MT
7	South Korea	589.2 MT
8	Islamic Republic of Iran	563.4 MT
9	Canada	540.8 MT
10	Saudi Arabia	527.2 MT
11	Indonesia	454.9 MT
12	Mexico	445.5 MT
13	Brazil	416.7 MT
14	South Africa	414.4 MT
15	Australia	392.4 MT
16	United Kingdom	371.1 MT
17	Turkey	338.8 MT
18	Italy	325.7 MT
19	Poland	293.1 MT
20	France	292.9 MT

There are fears that, as India becomes one of the leading emitters of carbon, its efforts to deal with the problem will be compromised by its concern not to disrupt familiar patterns of economic growth. There is already pressure on India & China by the developed countries on deadlines for controlling carbon emission. As China and India compete for oil and gas & new sources of energy, they will need additional resources and this could become a source of contention, if other countries have accepted plans for curtailing production of hydrocarbons. (Moran 2011; 74-80)

CLIMATE CHANGE AND REFUGEES

Climate Change has resulted in "climate change refugees". Human displacement occurring as a result of climate change is a comparatively recent development in comparison to the more traditional ideas associated with refugees, such as persecution based on race, religion, nationality, or membership.

Climate Refugees can be created in the following ways-

- Displacement of people due to changing sea levels- It is estimated by the IPCC indicate that by the end of this century the global sea level will rise somewhere between 28 and 43 centimeters as a result of thermal expansion and the melting of glaciers and ice caps (IPCC 2007a: 409). This is a conservative estimate. Actual indicators may vary from 40 to 150 centimeters within the same timeframe

(Black, 2008). The small island states are more likely to suffer disproportionate consequences especially in terms of land loss.

Many countries with low-lying coastal areas are similarly under threat. The Ganges-Brahmaputra-Meghna delta, which stretches from India and Bangladesh, to Nepal, China, and Bhutan, is home to approximately 129 million people. Rising sea levels, vast flooding, and resulting salt contamination of previously arable land has seen Bangladesh suffer more than most of the other countries.

- The previously vast rice fields and agricultural land, which provided vast resources and employment opportunities to the local communities, have been replaced with export-based shrimp farms due to salt contamination. IPCC has predicted that a rise of 45 centimeters would displace 5.5 million people and submerge over 10% of Bangladesh, the number of migrants will increase (IPCC 2001:569).

  This will add to India- Bangladesh conflict as India remains reluctant to absorb the big flow of climate refugees.

- Climate refugees also result from glacial Lake flooding. This is more prominent in the Himalayan region which has forced local people to migrate. It is estimated that by 2080, there will be between 50 million- 200 million displaced people owing to direct impact of climate change. (Williams 2008; 504-506) (Nicholls 2004; but cf. Black 2001: 7–8).

CLIMATE CHANGE AND THE GENDER DIMENSION

Natural Disasters and their Impact on Women in South Asia

An important aspect that needs to be understood is that Climate change is not a gender neutral phenomenon. As mentioned earlier,crisis, change and conflict affects women and men differently. Therefore there is an urgent need to understand that climate change attacks women in an entirely different as well as disastrous manner. Women and girls are less prepared to face the consequences of climate change on account of their low literacy levels,their economic status,low awareness levels about disaster management and less preparedness in terms of access to support systems,transportation and movement. Situation become worse when they are part of regions already vulnerable to climate change and gender inequality increases the vulnerabilty of women and girls to climate change. For example, 'natural' disasters, on average, kill more women at a younger age than men. In the Bangladesh cyclone of 1991,those women who died were four times more than that of men. It is the result of women's incapacity to receive early warning signals floated in public spaces,insufficient or no access to technology,cultural barriers which prohibit women to be informative and assertive about their rights and their dependance upon male members of their families to save them in times of crisis further limit their risk taking appetite.

It should be made clear that contrary to the common belief that women and girls are more vulnerable to disasters because of their lack of physical strength,the fact is that they suffer because of the social, cultural and psychological factors. In a study 'Gendered Nature of Natural Disasters by London School of Economics' (Neumayer and Plümper, 2007)explain that the large scale gender differences in mortality do not take place on account of physiological and biological differences between the sexes. There is a strong cultural dimension attached to the response towards disasters. An 'appropriate' behaviour is expected from men and women even in times of natural disasters. This can be explained with the help of the comparative study of the mortality among men and women when hurricane Mitch struck Central America with that of Tsunami striking Asia. More men were reported dead in comparison to women in the first case,because they undertook less precautionary measures than women in central America. In case of Tsunami,more women had fewer chances to survive Tsunami because they were not trained into swimming and climbing trees as men and boys were before Tsunami struck them. When men die,get lost or injure,then the entire responsibilty of the family comes to the women. In traditional societies,women are more vulnerable to violence when they move out alone,in search of basic amenities like food,fodder and water. Thus there should be a gendered understanding of the impact of climate change. (Otzelberger 2011; 5-14)

Climate Adaptation Strategies—In climate adaptation strategies,a gender blind approach works against women. The forced use of technology in the agricultural sector marginalises women and completely bypasses the traditional knowledge of women in the agricultural sector. Further, in post disaster scenario,the state reduces public expenditure and aims at downsizing. Women are the first to be laid off as they are not considered to be the bread winners in the patriachal set up of societies. The government spends less on food,health and education,thereby women have to compromise on their well -being in terms of food and health benefits making provisions for the rest of the family. Thus budgetary shifts of the government leads to gender inequalities.

Impact of Climate Change on Women in South Asia

The Report of the Inter governmental Panel on Climate Change(2014) has predicted that South Asia will experience overall warming and extreme weathers in near future. Food security,water availability will also be badly affected. It is already noted that large scale migration is taking place due to the impact of Climate Change in South Asia. It is important to note that the large scale migration is both a result of adverse conditions in places of residence as well as movement towards prosperous areas. Generally the menfolk migrate in search of better jobs and better incomes leaving behind women,children and the elderly. It should be noted that among those who are left behind,women are the worst sufferers because they are supposed to take care of the elderly and the children with no strong support structure. Thus in South Asia we have several Female Headed Households. The tragedy of these women is that they are forced to accept responsibility out of compulsion. They are inept,mostly uneducated and largely poor. Hence the terrible amalgamation of illiteracy, poverty and unskilled labour makes situations extremely challenging for them. (www.actionaid.org). Following are some of the prominent trends related to women in South Asia.

- ***Feminization of Poverty***: South Asia has seen a growth in women becoming poorer. Thus poverty is now having the face of a woman. In south Asia,there is a strong co- relation between women and informal work employment.(Doane, 2007).Statistics prove that 70% of the two million people suffering from poverty are women (ADB, 2009). Globalisation has made the situation more difficult for them and the social benefits also do not reach them.
- ***Feminisation of Agriculture***: Climate Change forces men to migrate to other places in search of better prospects Therefore,the entire responsibilty of land and livestock has comes to women. They are supposed to engage in tilling,cultivating,spraying pesticides and harvesting as well. The agricultural produce is to be sold by women in markets and they are also expected to cater the interests of the middlemen and agents for a good return. The most unfortunate part is that they do not even have ownership rights over the lands they cultivate. (Hassan Ahmed, 2014; vol. 5, No. 2)
- ***Feminisation of Labour:*** The ILO report of 2011 suggests that more than 80 % of women in South Asia are engaged in vulnerable employment. They mostly perform casual labor and are absorbed in the informal sector. Their working conditions are pitable and they hardly get decent wages. Ideally,as the level of development increases,there should be more participation of women in the formal sector,but in this region in the past 20 years there is merely 14%-15% of women in the informal sector. (www.in.boell.org; 2012)

- ***Victimisation of Women:*** The above mentioned trends in South Asia highlights the most painful trend which is present in all the trends discussed so far i.e., the victimisation of women. Women in general and in South Asia and Africa in particular, become victims in such situations. They are more vulnerable to sexual exploitation, economic exploitation,physical assault,fraud,cheating and trafficking among others. They are only made to work,but their work is not registered. They are farmers but not owners. They are workers without wages. It is a strange situation in that part of the world where women are revered as 'Devis'(Goddesses) but often treated as 'Dasis'.(Slaves)

Projection of Women as Agents and not Victims

It is essential to move beyond the popular discourse from treating women as merely victims. It is true they are victimised by the impact of climate change and the accompanied social,cultural prejudices against them. The study of women and climate change should not be limited only to the understanding of women as badly affected by climate change,it should progress to find out how women survive these changes and become agents of sustainable development. It is more important to celebrate their victorhood not their victimhood. The effort should be to acknowledge the role of women in sustaining the environment and moving towards sustainable development. The fresh perspective to look towards women can be as Agents,as survivors, as victors and not as victims of change and development. The western model of development does not offer all the answers to the ill effects of development. An alternative model of development in South Asia with Women as Agents of Sustainable development can be a good possibility to reinvigorate our roots and present at least an alternative discourse in International Politics.

Women as Agents of Change in South Asia

South Asia has been a conflict zone with heavy military deployment. Women have been subjected to immense trauma and pain as a consequence of conflicts. Climate Change has worsened the situation. However, the need of the hour is to move beyond the discourse of victimhood and highlight the agency of women. Conflict and post conflict societies women assume new roles and come out of their traditional households out of compulsion and necessity. They continue performing their roles of caregivers and nurturers but they also learn survival skills. In kashmir,which is a conflict zone, surviving means being more aware about the surroundings(knowing about strikes, curfew, explosion).Housewives get engaged in going to prison cells and detention centers searching for the 'disappeared,' Association of Parents of the Disappeared Persons' emerged out of the efforts of a woman named Parveena Ahangar,whose search for her disappeared son led to the formation of this organisation in Kashmir. In Nepal,when villages were completely wiped off of men by the Maoist,women bravely formed the backbone of the agricultural economy of these villages and successfully undertook ploughing and other farming activities (traditionally associated with men). In Sri Lanka, war pushed women into public arena. They learned new skills -technical,commercial, professional. Women undertook lagoon fishing, food processing and marketing. In LTTE controlled areas,they emerged as brave entrepreneurs owning fishing boats and also employing men. They even challenged caste hierarchies, and social practices. In Afghanistan,in post Taliban rule,women started underground schools. They lived together in all women families(as their men were killed or lost) in compounds headed by strong matriachs. Not only this,women in Sri Lanka and Afghanistan have also indulged in demining their areas. This is a very courageous and risky task feared by men. Women have taken up the responsibility to do this, irrespective of the risks and the social pressures involved in the entire process of becoming agents of change. (Manchanda, 2005; 4738-39)

Women have proven their agency in post conflict scenarios. In the same vein, they hold promise in dealing with the challenges of climate change through their traditional knowledge yet progressive actions.

The next part proposes the indigenous model of development that should be followed by the South Asians with women as agents of this model of development.

The Indigenous Model of Development in South Asia

South Asia has a long history of colonialism and its damages. The colonial mindset and the slavish mentality have been responsible for imitation of the west. This tendency of mindless aping the west has percolated all aspects of living. The academia and its research is also not devoid of such Eurocentric study of indigenous problems. It is true that climate Change is affecting the world, yet South Asia is more badly affected than the developed countries. It is already battling its specific problems of overpopulation and under utilization of resources, therefore it requires a model of development which is progressive yet rooted in its cultural background. A model which is an amalgamation of economic progress and peaceful living. As Dr. S.P. Varma wrote in his popular book 'Modern Political Theory' way back in 1970s "As long as development is perceived in terms of movement towards mass consumption society, the poor nations, rather than seeking to emulate the rich have to redefine their goals of development".The goal of development should be harmonious and peaceful living. Goal can be achieved with the active participation of all stakeholders in the process of development. Women, in particular have proved to be agents of peaceful development. They have acted as responsible custodians of nature in the March towards development. In South Asia,women have been trained into living peacefully with nature rather than using nature for their convenience. Traditionally, they support a harmonious ecosystem. Thus, Environmental Ethics is not an academic pursuit, rather a way of life for the people of this region and women have beautifully undertaken the responsibility to nurture the natural world.

Environmental Ethics and Its Importance in South Asia

The environment is made up of human and non-human resources. Both are valuable to the environment,the west believes that the non-human resources hold instrumental value to the Humans, the east find the non-human resources like natural resources and wildlife of intrinsic value. This is the main point of difference between the West and the East on the issue of combating Climate Change. One holds the environment important to serve itself, hence anthropocentric approach to the environment (protecting the natural resources because they sustain human beings), the value of these resources is therefore limited to the instrumental value attached to them**.** The east does not consider these natural resources and wildlife as instrumental; they are revered on account of their intrinsic value. The difference is substantial enough to prove that South Asia is a region which does not have to learn from the west, it in fact has to tutor the developed countries in sustainable living. Dharmic Environmentalism, stresses upon the beautiful relation between environment and human beings. Trees, rivers, mountains are not worshipped because of their use value, they are revered because they are so. The 'Sacredness' attached to the natural resources prohibit them to be destroyed or recklessly used. This promotes Man -Nature Bonhomie.

Understanding Hinduism and Ecology

Hinduism is a prominent religion of South Asia as well as of the world. The influence of Hinduism is widespread in the entire region. The religious texts and scriptures are authentic sources of knowledge. They stress the importance of ***Rta*** or the great cosmic order, the order means the harmonious relation between man and nature. The symphony of this order brings happiness and peace to all, but any imbalance in it is supposed to lead to destruction. This religion teaches that human beings are part of an order which is greater than them. There is a specific role assigned to all the parts in that order. There is a natural cycle in the universe. Sri Krishna's exposition to Arjuna in this verse explains :He who does not follow this ever-revolving natural cycle, who is of sinful life, rejoicing in the senses, he lives in vain, O Arjuna! (Paranjape; 13) . Thus human beings are not supposed to utilize the resources in a selfish manner, but in a sustainable manner. They should follow the path of Dharma. The concept of Dharma stands for righteousness, and duty. It stands to uphold the order. Importance of Panchmahabhutaas in Hinduism. The five "Tatvas"-elements (Akash-Space, Vayu-Air, Jal-Water,,Prithvi-Earth, Agni-Fire) are interconnected to each other and are connected to the human beings, as the human body is composed of them. Therefore Nature and human beings are inseparable from each other. Human beings should not strive for personal fulfillment but universal peace.

Hinduism as a Religion of Harmonious Living with Nature

Hinduism believes in the omnipresence of God. Since God resides everywhere and in everyone, therefore one must practice love and compassion towards all. The concept of Ahimsa is well rooted in Hinduism. Hinduism worships rivers, lakes, animals, birds, reptiles, forests,trees,bushes and grass. The concept of sacred groves is one such example. India is home to various sacred groves in Uttrakhand, Kerala, Rajasthan among others. Sacred groves comprise of trees like Deodar (Cedrus deodara), considered to be the "abode of Gods" Sal (Shorea robusta), Rudraksha (Elacocarpus species), Bael (Aegle marmelos), and Ashok (Saraca asoca), and kadam (Anthocephalus kadamba), Pipal (Ficus religiosa), Neem (Azadirachta indica), Banyan tree (Ficus benghalensis): native to India, Mango tree (Mangifera indica) and bushes like, Basil(Ocimum basilicum & amp; Ocimum minimum) (native to India and Iran) and grass like Doob or Durva (cynodon dactylon) etc. (www.ecosensorium.org 2010; 02)

Women take special care of the sacred groves. They nurture them with utmost love and devotion because they believe that they are the abode of Gods. They are devotees in this sacred space, hence trees are not looked as trunk and branches put together for bark, leaves and fruits. Trees are Gods to be worshipped, protected and loved.

Hindu Way of Life and The Role of Women

A certain way of life where in a typical hindu family, women cook the first chapati of the day for the cow, a chapati is saved for the dog, earthen water vessels are hung on trees for the birds, women make beautiful kolams (with rice flours) in the morning outside homes to feed the ants and birds. The 'arti' in the evening and the worship of the Tulsi plant are rituals which make women the link between nature and mankind. Many more daily rituals of Hinduism are important steps to reduce the impact of climate change-to put more technically, reduce our carbon footprints.

Women in South Asia can offer guidance to the ***Eco feminists*** who espouse the same cause,although at a different level. Eco feminism believes in the relationship between social and environmental issues. The environment as well as women are subjected to the exploitation by a common source of patriachal culture. Thus it is radical in nature. It offers a solution in the form of harmonious living with nature with the elimination of all forms of exploitation. (www.voicesforbiodiversity.org)

Successful Case Studies of Women as the Protectors of Nature

1. **Bishnois of Rajasthan:** The religion of Bishnois is based upon environment protection. Bishnoi is made up of Bees (20) and noi (9), means the nine rules offered by the founder of this religion, Guru Jambeshwar. These rules are related to wildlife, nature protection as well as social behavior.A true Bishnoi is the guardian of nature and wildlife. This belief system considers trees as sacred and animals as their own children. Endangered species like Blackbucks and Chinkara deer find a safe territory amongst Bishnoi communities. This community has made environment protection their religion; it is a part of their daily lives. Everyday practice like keeping water vessels and hanging water pots for animals and birds is a common sight in the Bishnoi area. Women breast feed the young ones of blackbucks along with their own. This is a rarity found only among the Bishnois. Research fails to find another example of dedication and commitment towards conservation of environment. The essence of religion as harbored by Smt. Amrita Devi (the woman who led the first movement of tree hugging in India as a protest against the order of the Maharaja of Jodhpur for cutting the Khejri trees, this variety of tree is revered by the Bishnois) still lives on.Amrita Devi sacrificed her life for the sake of protecting the trees, and till date Bishnois are carrying forward this legacy of Smt. Amrita Devi. (www.thebeeterindia.com)· Another notable feature of this special community is that it has become active towards protecting the wildlife from poaching, hence strict patrolling by the Bishnois have been a detriment to the poachers. They value the life of the sacred animals more than their own and do not hesitate to sacrifice it for the sake of the animals.The hedonistic, atomistic understanding of preservation of nature for its instrumental value supported by the most erudite academicians stand defeated in front of the live example of such a wonderful community. Bishnois prove that humans do not interfere in Environment; on the contrary they conserve and nurture it. As Ann Gold writes about the village communities in Rajasthan (2001) where individual, communal, cosmic and geo-physical interdependencies are systematically ordered and perpetually negotiated in this locally posed ecologies. (Jain, 2016)
2. **The fight of the Dongria Konds in Orissa:** In 2009, James Cameron movie "Avatar" became a block buster. Everyone commended the

resistance of Na'vi. The fantasy projected in this movie has a reality, the Dongria Kond tribe of Orissa, a tribe which has worshipped Nature in all its forms, for generations. The niyamgiri hills locally called the Niyam raja -is revered by the tribe, which is self-sufficient. In their own words it is their Mother, their religion. This hill is a storehouse of about $2 billion worth of bauxite. Bauxite is a boon for the tribal. During rainfall Bauxite and other minerals holds water in suspension and throughout the year, slowly keep supplying this water in the form of streams to the tribals below. Anil Agarwal, the London based owner of world renowned Vedanta, planned to open an aluminium open cast mine in the area of Niyamgiri. This mine would have abrogated the basic right to life of these people, right to life, right to food and right to natural resources. Dongria Konds,among them a large number of women, started their protest to this mine because they believe that the mine will not only destroy them, more important than that hurt the Niyamgiri hills-their God. Therefore they gave a strong resistance to it and Vedanta group had to ultimately stop its work (W'ell lose our soul. Niyamgiri is our soul). (www.survivalinternation.org) Anil Agarwal, the billionaire had to accept that Orissa project was a mistake. Orissa and Andhra Pradesh house large areas of bauxite capped mountains and tribals in these areas have strongly resisted any attack on these mountains. The determination of the dongria Konds simple tribe proves that women do not have to depend on others to fight for their rights, sometimes all that matters is just courage and faith.

3. **Women as participants in environmental management in Sri Lanka:** In the Kelegama, Labunoruwa and Muriyakadawala, which fall under the dry zone of Sri Lanka, women have maintained ecological sustainability. This is a 'hydraullic civilisation' developed with traditional knowledge and sound resource management. In Sri Lanka, as in India,a forest resource based culture forms the basis of ecological sustainibility. Women take care of household livelihoods as well as subsistence. As gendered roles are pre-decided, women interact with the ecosystem more frequently than men. They are food gatherers, on a daily basis. Forest is their daily point of contact. Since forest are directly related to the subsistence,hence women are directly affected by deforestsation. As patriachal set up permits women to interact with the ecosystem in a more integrated manner,women become the repository of traditional wisdom,knowledge and thus best placed to provide ecological sustainability.

This is a dry zone,reservoirs are built to store water by the villagers to receive water supply throughout the year,even in times of drought. Women's traditional knowledge has been instrumental in the preservation of forests, which are very important in the generation and storage of water. Forests are instrumental in controlling heat,trapping clouds and generating rain. Women's knowledge has helped in the conservation of water.

Women's knowledge of the food plants,food culture has helped in the evolution of a food system based on a judicious use of food resources. They have helped in regeneration,conservation,produ ction of important resources. They stress the acknowledgement of their traditional knowledge and opposes mechanized farming. (Wickramasinghe, 2003)

CONCLUSION

The developing countries have a tendency to look towards the west for development, however in the changed scenario, it is not a compulsion. It is a choice of a particular country or a region. The Academic circles in India have long been mesmerized with the discourses of the West, created by the west and popularized by the West. In contemporary times, there is a serious requirement of an alternative discourse. Environmental Politics cannot ignore the importance of these alternative discourses. South Asia offers an alternative discourse on environment protection. It provides a rich framework for the development of a new model. Based upon sustainable living this region presents interesting research material for investigating the Man –Nature Bonhomie. South Asia provides an example to the world that conservation of nature is inbuilt in its culture. Sustainable living is a part of its religious environment. The need of the hour is to change the focus of research towards the lesser known facts about this region. It is equally important to carefully address the specific problems of this region; therefore it is imperative for the national governments to balance between immediate growth and sustainability in the most prudent manner. The future of South Asia depends upon the wisdom of the policy makers of this region. The success of its model of sustainable development depends upon the intelligence of the academic fraternity.

The role of women in protecting the environment is also to be understood and presented in a new light. The importance of women as agents of change should be recognized in the academia as well as in the environmental policy making process. It is long enough to project women as victims of change. They are the victims of change,but they are also an agency to bring change. The commendable work of women in south Asia in the protection of the environment needs to be highlighted. Their efforts for the conservation of nature should become an interesting source of research for those working on alternative paradigms of development. The beginning of a dialogue,discussion can open a new arena of understanding nature through women and the inter-relationship between the two can prepare a guideline for future course of action.

REFERENCES

1. South Asia Environment Outlook2014, UNEP Regional Office for Asia and the Pacific (UNEP/ROAP) ISBN : 978-92- 807-3402- 7, Job number: DEW/1823/NA
2. https://www.theguardian.com/global-development/2012/apr/10/slum-cities-south-asiaplanning.
3. http://www.worldbank.org/en/region/sar/publication/urbanization-south- asia-cities
4. Dev Mahendra "Climate Change, Rural Livelihoods and Agriculture(focus on Food Security)in Asia Pacific Region", Indira Gandhi Institute of Development Research, 2011,pp.1-11. Available at:www.igidr.ac.in/pdf/publication/WP-2011-014.pdf
5. Moran Daniel" Climate Change and National Security" Georgetown University Press, 2011 pp. 74-80
6. Williams Angela "Turning the Tide: Recognizing Climate Change Refugees in International Law", Baldy Centre for Law and Social Policy, 2008, pp. 504-506.Also available at:http://muechner-fluechtlingsrat.de/uploads/Williams-turning-tide.pdf
7. Otzelberger, Agnes"Gender Responsive Strategies on Climate Change: Recent Progress And Ways Forward For Donors"Institute of Defence Studies, June 2011,pp-5,14 8.
8. http://www.actionaid.org/sites/files/actionaid/climate_change_migration_in_south_asia_web_version.pdf
9. Hassan,Syeda;Ahmad,Khalil,"Globalization: Feminization of Poverty and need for Gender Responsive Social Protection in Pakistan". Also available athttp://pu.edu.pk/images/journal/studies/PDF-FILES/Artical-3_v15_no2_14.pdf
10. http://www.fftc.agnet.org/library.php?func=view&id=20110725164020
11. https://in.boell.org/2012/05/25/informalisation-womens-work-india
12. Manchanda,Rita,"Women's Agency in Peace Building:Gender Relations in Post Conflict Reconstruction", Economic and Poltical Weekly,Vol. No 44/45,October 29-November 4,2005, pp-4738-4739
13. Paranjape, Mark and "Hindu Ecology" in "Religion and Ecology" Sokka Gakai International quarterly magazine, pp-13

14. http://ww1.ecosensorium.org/2010/02/what-are-sacred-groves-and-why-are-they-html
15. http://voicesforbiodiversity.org/articles/the-growing-importance-of-ecofeminism
16. https://www.thebetterindia.com/5621/the-land-of-the-bishnois-where-conservation-of-wildlife-is-a-religion/
17. Jain,Pankaj" Dharma and Ecology of Hindu Communities: Sustenance and Sustainability", Routledge,2016
18. http://www.survivalinternational.org/tribes/dongria
19. Wickramasinghe,Anoja,"Gender and Ecological Sustainability: The Traditions and Wisdom Of The local Communities in A Dry Zone, Sri Lanka" in, (Ed)Krishna, Sumi"Livelihood and Gender: Equity in Community Resource Management" SAGE Publications Pvt. Ltd,2003

Chapter 7

Sustainable Development Goals: Integration of Information and Communication Technologies (ICT) to Deliver Sustainable Development Goals Myth Vs Evidence

Sarfraz H. Khawaja and Akbar Ayesha

ABSTRACT

Smart policy designs and ideas are best vehicles to address gaps in ensuring social cohesion, poverty alleviation, promoting education and better health care standards. SDGs is an avenue to deal with main challenges of development, democracy and defense. The novel strategies can be solicited to improve better policy making and decisions to foster efforts for achieving SDGs. Governance in the wake of 21st century has shifted its focus to make the system more transparent, accountable and deliverable by integrating the concepts of quantifying performances and service delivery. New public management concepts provide impetus and breeding grounds to base decisions on the data and evidence for eliciting the concept of evidence based policymaking. Achievement of SDGs through integration of ICT are in incredible relationship to bridge the gaps in service delivery. The breakthrough in technology can be exploited with proper integration of policy interventions and find loopholes in the administrative system. The transformation of manual records to the digitized "Dashboards" would provide the reflection of over all performances to target the areas of underperformance or to highlight the efficient services. Pakistan is on its way to embed smart policy designs in wider sectors of health, education and land record management. Achievement of the Sustainable development goals is the hallmark of development. Integrated analysis of polices is required to integrate ICT models which enables policy planners to come up with the intervention for achieving SDGs.

Keywords: ICT, Evidence Based Policy, smart policy designs, SDGs.

INTRODUCTION

The Sustainable Development Goals is an agenda of enlightenment for the civil societies to be pursued by and jointly implemented by the states. The literature reviewed in this context reveals that the capacity of many countries mostly in Africa, Asia and Latin America lacks necessary mechanism. There have been several countries with impressive economic growth but poor Human Development Index. Pakistan is one such country suffering because of poor governance. The good governance can neither be properly understood nor genuinely implemented without creating a society which is knowledge friendly and not averse to new challenges. Knowledge is the new source of wealth and is functional at two critical aspects of competitive advantage which is productivity and innovation – when knowledge is applied to what we already know, we call it productivity and when it is applied to new and different things we call it innovation. The Sustainable Development goals need these two pillars of productivity and innovation for achieving SDGs.

In Pakistan good governance is one key for focal areas to consider in the social policy of Pakistan. The issues of governance range from sectoral problems to the systemic and structural problems of the country. Both the sectoral and systemic challenges are interlinked in terms of their impact on the performance in the provisions of public welfare.

Smart policy designs are one of the best vehicles to address the gaps in ensuring social cohesion, poverty alleviation, promoting education and better health care standards. SDGs is an avenue to deal with main challenges of development, democracy and security. The novel strategies can be soliciting to improve grounds for the better policy making and decisions to foster efforts for achieving SDGs. Governance in the wake of 21st century has shifted its focus to make the system more transparent, accountable and deliverable by integrating the concepts of quantifying performances and service delivery. New public management concepts provide impetus and breeding grounds to base decisions on the data and evidence for eliciting the concept of evidence based policymaking.

On 1 January 2016, the 17 Sustainable Development Goals (SDGs) of the 2030 Agenda for Sustainable Development — were adopted by world leaders at an historic UN Summit. Over the next fifteen years, with these new Goals that universally apply to all, countries will mobilize efforts to end all forms of poverty, fight inequalities and tackle climate change, while ensuring that no one is left behind. The SDGs, also known as Global Goals, build on the success of the Millennium Development Goals (MDGs). The new Goals are unique in that they call for action by all countries, poor, rich and middle-income to promote prosperity while protecting the planet. They recognize that ending poverty must go hand-in-hand with strategies that build economic growth and address a range of social needs including education, health, social protection, and job opportunities, while tackling climate change and environmental protection. (www.un.org/sustainable development)

THE CASE OF PAKISTAN

Pakistan to achieve agenda of SDGs has made national framework to establish a link between modern technology and resource mobilization. The major initiative includes an integration of ICT to bridge the gaps in service delivery. The breakthrough in technology can be exploited with proper integration of policy interventions and by finding loopholes in the administrative system. The transformation of manual records to the digitized "Dashboards" would provide the reflection of over all performances to target the areas of under performance or to highlight the efficient services. The integrated model of ICT offers capability of designing administrative structures with effective techniques to ensure SDGs.

Sustainable development goals are the hallmark of development. Integrated analysis of polices is required to integrate ICT models which enables policy planners to come up with the intervention for achieving SDGs. The government stakeholders with these structural changes can chalk out plans to involve local government actors as they are in close contact with the community. These reforms aim at better communicating the needs of society, and country at large to deliver sustainable development goals. This activity will stimulate more options for cross-sector comparisons to improve the indicators of all SDGs. Performance driven culture endeavors an incredible relationship between service sector outcomes and smart policy options.

SIGNIFICANCE OF THE STUDY

After prolonged deliberations Pakistan is on the road of development and to embed smart policy designs in wider sectors of health, education and land record management. Evidence based policy making has brought a paradigm shift for reforms, policy formulation and implementation. Pakistan is moving from a security state to a welfare state despite several security challenges both internal and external. Human security signifies a varied approach in different regions depending upon the leadership and policy priorities. The analysis of the Human Development Index (HDI) shows that the link between human development and economic growth is not automatic. The link has to be created consciously by forward-looking policies and strategies by national governments and these policies should ensure that the goals relating to health and education would be taken up for policy evidence to integrate ICT for the improvement in service delivery.

METHODOLOGY

Multipronged research strategies have been employed to gather empirical data to validate statements. The research is broadly based on descriptive and theoretical analysis to provide sound understanding of the information. Qualitative research methods provided the base for content analysis of historical and contemporary literature. In addition, websites, international and national publications, United Nations' report and the analytical analysis of start-ups in Pakistan provided a great repository of data to build analysis on the status of SDGs.

Objective:

- Role of smart solutions to track students' and teachers' performance
- How smart solutions can effectively maintain health administration

ROLE OF SMART SOLUTIONS TO TRACK STUDENTS' AND TEACHERS' PERFORMANCE

A Reflection of United Nations on Goal-4 EDUCATION Quality Education – Why It Matters

Ensure inclusive and quality education for all and promote lifelong learning:

Education is the key that allows as many other Sustainable Development Goals (SDGs) to be achieved. When people are able to get quality education they can break from the cycle of poverty.

Education, therefore, helps to reduce inequalities and to reach gender equality. It also empowers people everywhere to live more healthy and sustainable lives. Education is also crucial to fostering tolerance between people and contributes to more peaceful societies.

Facts and Figures

Enrolment in primary education in developing countries has reached 91 per cent but 57 million children in developing countries remain out of school, more than half of the children not enrolled in school live in Sub-Saharan Africa. An estimated 50 per cent of out-of-school children of primary school age live in conflict-affected areas.

Table 1: A Critical Scenario of Education in Pakistan

Indicators	*Target*	*2015*	*2017*	*Status*	*Challenges-2030 (Gaps to be filled)*
Literacy Rate	100	58%	58%	0 off- track	42%
Gross Enrolment Rate	100	90%	87%	-3% off-track	13%
Net Enrolment Rate	100	57%	54%	-3% off-track	46%
Urban_ Rural Gulf					
Urban	100	74%	74%	0 off- track	26%
Rural	100	49%	49 %	0 off- track	51%
Male Literacy Rate	100	69%	81%	+12% off-track	19%
Female Literacy Rate	100	45%	68%	+23 % off-track	32%

1.1-Time Series Analysis of SDGs in Pakistan

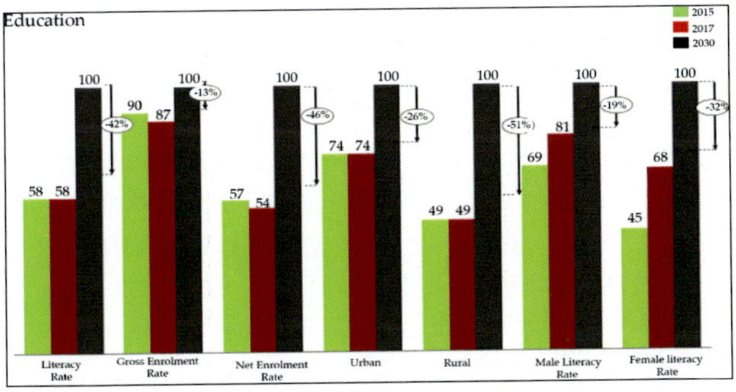

Source: Economic Survey of Pakistan (2015-2017)

Graph 1

Graph 1 indicates that literacy in Pakistan did not change during 2015-2017, whereas Net Enrolment Rate shows 3 percent decline which was recorded 57 in 2015 and now it stands at 54 percent. Gross Enrolment Rate also reveals declining trend of 3 per cent which was recorded in 2015 at 90 percent and now it stands at 87 per cent. Urban and Rural literacy ratio also stands still by showing "O" improvement. Notwithstanding gender disparities between male and female, the gap is being filled by improvement of 12 and 23 percent respectively.

On SDGs index Pakistan ranks at 149/188.7(www.sdgindex.org). This demonstrates evidence of a learning crisis when only 55per cent of school-going children are able to read stories in Urdu, Sindhi, and Pashto. According to a 2015 Annual Status of Education Report (ASER), 49 percent comprehend English sentences and 50per cent are able to solve two-digit math divisions. With an education challenge of such magnitude, it makes good sense shifting the national agenda to achieving learning outcomes, thereby sharing focus with universal primary enrolment goals. However, this realization has taken a quarter of a century to materialize within global education agendas. 52 percent of government primary schools are without toilets and 40 per cent don't even have drinking water. The ASER 2015 survey found 20 percent of children aged between six and 16 were out-of-school – and the remaining 80 per cent enrolled in schools not learning much given the inadequate learning methods used. (aserpakistan.org/index.php)

Education leads to increase in economic avenues and contributes in nurturing a healthy social sector. In Pakistan underspending on educational budget and lack of teacher's training in adopting new pedagogical techniques lead to poor learning outcomes among children. Right to information about numeracy in education should be made public to sensitize the government and private stakeholders to bridge the gaps by setting standards for tracking performance. Irony is, provinces with lesser autonomy of decision-making and low budgetary allocations cannot transform the educational system fully. Gender disparity, urban-rural divide in educational opportunities and exclusion of marginalized people fuse the good prospects of achieving goal-4. The status of education in Pakistan needs serious deliberations by engaging bureaucracy, civil society, state & non-state actors and international agencies for the development of policy insights. Role of decentralization of education to explore potential prospects, fiscal support and modalities also require critical analysis as the local actors are catalyst of change.

A Road Towards a Literate Pakistan

Pakistan Vision 2025(2014) of Ministry of Planning Development & Reform in fulfilling the aspirations of development has started technology led service sector as major move towards accomplishments of SDGs. It involves a rapid scaling-up of investments in education, health and social development, generating jobs and prospects for the youth bulge.

A Catalyst of ICT–Punjab Information Technology Board

Punjab Information Technology Board in realization of technology for the management of administration have introduced a myriad of projects to generate evidence to form decisions. Integration of ICT for maneuvering governance strategies have been considered a drive of reforms to set targets for achieving Goal-4 of SDGs.

E-Learning

E-learning enables teachers to deliver lectures with the help of sophisticated digital devices. Android tablets have been handed over to teachers equipped with semi-smart TVs. It helps teachers in developing curriculum, plan training activities and enable students to have online access to study material by using user-friendly interface. Offline availability of content has also been assured to give full-assess to course content to students. eLearning Punjab aims to facilitate and encourage the use of educational by providing digitized content, relevant supplementary resources, and online assessment tools for students and educators. It targets to address technological gaps, address adoption issues and correct content weakness.

Figure 1 : E-Learning Models

Source: e-learning Punjab Website

Smart Monitoring of Schools: This initiative is part of the Open Data Strategy of Government of the Punjab. The Government of Punjab employs 950 monitoring officers across the Province. These officers act as an independent school monitoring layer for the Government, and are tasked with making spot-visits to its 52,695 public schools across the Punjab, each month. During these spot-visits, data is collected on teacher presence, student enrolment and attendance, as well as availability of facilities such as clean drinking water, electricity, and toilets. Public access to this data is now available online for free. Provincial and district level summaries are computed automatically based on data from actual forms submitted by the monitoring officers. Website visitors can navigate their way down to the actual forms submitted by the monitoring officers.

Literacy and Numeracy Drive: is a project of PITB to evaluate the performance of schools. The program evaluates the performance based on the criterion of pre-designed Students Learning Objectives (SLOs). The designs aim at defining the criterion of structuring student's curriculum. Monitoring and Evaluation Assistants were given devices with installation of applications with a series of tests comprising of multiple choice questions based on SLOs. An average of 49,000(www.pitb.gov.pk) public schools is visited on a monthly basis, with approximately 2,400 visits each day. It means that around 14,000 students are assessed every day.

Data Driven Insights for Policy Designs: Centre for Eco- nomic Research in Pakistan (CERP) has introduced a magnificent mechanism of data driven insights for policy design in operating the administrative system. This ignites an approach of evidence based policy making in Pakistan. In an attempt to revitalize Pakistan's overall fabric with integration of technology, CERP started a project in 2002 with the name of LEAPS (Learning and Educational Achievement in Punjab Schools). The LEAPS Project investigates the changing educational landscape in Pakistan to synthesize findings and inform policy debates on education in the country. Against all odds of poor data on education, CERP advices, government should provide information for better decision making and accountability, increasing access to schooling, reform failing government schools, reform teacher compensation and government should act as an innovator. These changes represent an opportunity and a challenge for educational policy in the country(www.cerp.org.pk).

Online College Admissions System (OCAS) is built to provide easy, convenient, efficient and reliable access to students for admissions in the government colleges for higher level of education. The solution is based upon a collaborative relationship between Higher Education Department, Nine Boards of Intermediate and Secondary Education (BISE) of Punjab, Bank of Punjab and Punjab Information Technology Board (PITB)(www.pitb.gov.pk)

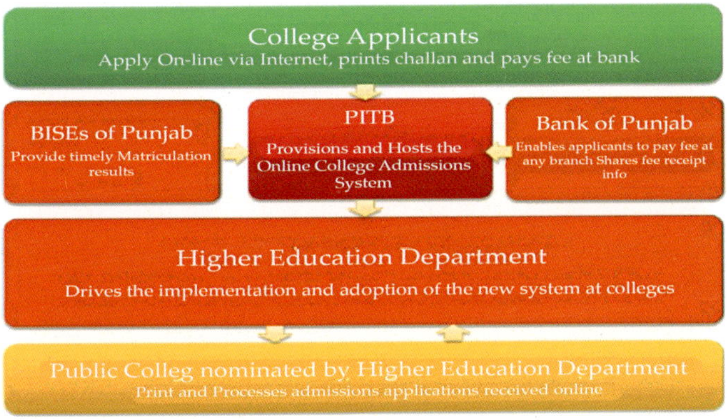

Figure 2: Public Colleges nominated by Higher Education Department:

Source: PITB Website

'The major contribution of OCAS towards facilitating students and their families in the admission process is saving time and cost. Whoever has access to the internet can apply to any number of colleges, anytime, from anywhere and initiate application process for college admissions across Punjab. From what was the submission of applications – taking hours and days for each segment of the application process – the same can be done in minutes and not requiring physical visits to colleges' (Khawaja 2017).

ILO & ICT provides an excellent opportunity for the teachers to adopt from several teaching styles which were previously confined to a few. It gives countries an opportunity to achieve the International Labour Organization (ILO) goals of Education for All (EFA), Skills for All (SFA) and Lifelong Learning for All (LLA).(Khan 2017) Though there is no denial that the modern technologies have helped in developing new ways and models to get paramount success in the teaching-learning process but have added challenges as well. Toffler said in 1991, "The illiterate of the 21st century, will not be those who cannot read and write, but those who cannot learn, unlearn, and relearn", and he was very true. A whole generation of those teachers who never touched a computer was forced to unlearn and relearn. Some did and some not, so is the success rate of inclusion and optimization of ICT in education. (Khan 2017)

Conclusion

The use of ICT in efficient ways can better mobilize the resources to achieve sustainable development not only in education but to achieve rest of the goals of SDGs. Pakistan with reforms should turn around educational and health metrics by 2030 to achieve SDGs and to stay abreast with the international community. Public-private partnership, integration of technology and adoption of international best pedagogical techniques can give a kick start to provide a solid pitch to play smartly.

Recommendations

- A national framework for policy can explore many avenues for awareness of excellence by providing teacher's training, monitoring and evaluation mechanism with smart and evidence based policy designs.
- Integration of ICT will not only generate evidence to inform policy and decisions, it will channel international funding and public financing with transparent information.
- The minimum influence of bureaucratic inertia and more autonomy to provinces with equitable budgetary allocation can provide legal umbrella to take head start initiatives.
- Government at all levels should design strategies by engaging civil society, political parties, and international donor agencies to make the system more accountable and transparent. Increasing the capacity of local authorities to implement plans can strengthen the performance of service delivery of education.

SMART SOLUTIONS FOR EFFECTIVE HEALTH ADMINISTRATION

A Reflection of United Nations on Goal-3 HEALTH

Health & Wellbeing - Why It Matters

To ensure healthy lives and promote well-being for all at all ages,

Pakistan has come a long way, but it still has a longer way to go. Real progress means achieving universal health coverage; making essential medicines and vaccines affordable; ensuring that women have full access to sexual and reproductive health care; and ending all preventable deaths of children. Inequalities in health care access persist. More than six million children still die before their fifth birthday each year, and only half of all women in developing regions have access to the health care they need. Epidemics like HIV/ AIDS thrive where fear and discrimination limit people's ability to receive the services they need to live healthy and productive lives(www.un.org/sustainabledevelopment).

HealthCare in Pakistan

According to UNICEF, despite significant improvements over the past two decades, Pakistan ranks towards the bottom among other countries when it comes to infant and neonatal mortality. This is why 44 percent of all children are stunted and 9.6 million experience chronic nutrition deprivation. Compounding the nation's state of poor health, for example, is the toll taken by pneumonia killing approximately 92,000 children annually. Pakistan's ranking in the Maternal Mortality Ratio Index has slipped from 147 in 2014 to 149 in 2015, recording a staggering 276 deaths per 100,000 births

Goal 3: Good health and well-being

- Targets: end communicable diseases, achieve universal health coverage, and provide access to safe and effective medicines and vaccines by 2030
- Pakistan is ranked 149th out of 179 countries in 2015 on the Maternal Mortality
- Ratio Index
- New healthcare blueprint needs increased funding, efficient cross-sector linkages

According to the DAWN's (daily newspaper-www.dawn.com) special coverage on SDGs previous 40 MDG indicators precludes Pakistan's ability to achieve the SDGs. The therapy being proposed is that countries should identify goals based on their resources and try to achieve them rather than opt for a one-size fits all framework. However, they are yet another articulation of aspirations reflected in development terminology crafted in the halls of the UN and subscribed to by governments without always comprehending the level of effort and resources required.

Critical Reflections

The good indicators of health reflect good socio economic standing, parents' education, adequate nourishment and good health of mother and child. Pakistan reveals poor statistics of health alluded to disproportionate relationship between high population and resources. The UN's agenda tables the initiatives in a Global perspective. However, the subjective ground realities of each country abound with reasons for not articulating the aspirations of SDGs. Poverty is one of the crucial factors, according to the Leadership for Environment and Development

(LEAD - Policy Think Tank) 21.04% population is below poverty line $1.25 whereas, 60.19% is below $2. This indicates that lack of accessibility factor directly influences high rate of maternal, neonatal and infant mortalities due to the malnourishment which reveals poor indication of development.

Table 2: Time Series Analysis of SDGs in Pakistan

Indicators	*Target*	*2015*	*2017*	*Status*	*Challenges - 2030 (Gaps to be Filled)*
Infant Mortality Rate (IMR)	As low as 12 *deaths* per 1,000 live births	66/1000	62/1000	Off-track	< 50/1000
Maternal Mortality Ratio	Less than 70 per 100 000 live births.	178/ 100000	170/ 100000	Off-track	< 100/ 100000
Stunted Children	100%	44%	44%	Off-track	56%
AIDS	100%	45990	132,000	Off-track	82% (esti-mated)
Polio	100%	54 cases	2 cases	On-Track	2%

Source: Economic Survey of Pak. (2015-2017), IRIS reports, 2017, Dawn Newspaper, 2017

Health profile of Pakistan is characterized by high population growth. Despite a significant investment in health sector over the years, the rate of progress is lower than most other countries. Pakistan has only registered a 10-point reduction in stunting prevalence during this period, from a 54.5% (Inter-Agency regional analysis Network-2017)stunting rate measured in 1990-91 to 44%((Inter-Agency regional analysis Network-2017)) measured in 2013 (Bhutto 2017). In Pakistan, malnutrition is widespread among all ages, and progress to address social determinants over the last several decades had been very slow. According to the National Nutrition Survey 2011, one-third of all children are underweight, nearly 44 percent are stunted, 15 percent are wasted, half of them are anemic and almost one-third of these children have iron deficiency anemia. These rates have hardly changed over two decades according to the findings of a maternal and child nutrition study group published by Lancet in 2013. The disease pattern is heavily dominated by malnutrition and poor dietary practices. SDGs is a national commitment to improve health outcomes. All federal and provincial/area governments are developing, implementing and monitoring health sector strategic framework to achieve health related SDGs (Economic Survey_2016-2017). Nearly 132,000 (Junaidi 2017) people in the country are living with HIV/AIDS, but only 15,370 are registered with 21 HIV treatment centers.

A critical link exists between planning process, planners, available resources, political will, civil provincial rivalries, and national and international funding. This creates a vicious cycle where citizens do not receive services and state due to the incapacity of administrative reforms is less willing to strengthen the link between good governance and citizens' satisfaction. Pakistan is a disaster prone state and climate change adds more challenges to tackle endemics and epidemics. Polio, Tuberculosis, Hepatitis A, B, C, malnourishment, stunting, and AIDS are critical health challenges for governance to achieve sustainable development goals.

Reforms towards SDGs Targets

Technologically led solutions to the problems ignites an orientations towards smart monitoring designs and coherent polices for ensuring compliance. Government of Punjab in realization of achieving SDGs has revitalized the health vision with integrated models of ICT to track distortions in the service delivery. PITB offered IT centric solutions with the help of World Bank to consolidate information to maintain supply chain of drugs, equipment and availability of human resources. These initiatives would be implemented across Pakistan after receiving successful outcomes in a due process of time.

Electronic Medical Record & Hospital Information Management System

- BHU Clustering
- Ensuring Health Service Standards through MEA Health
- Maintenance and Cleanliness System
- Health Watch
- Medicine Inventory Management System
- VET Inspect
- Monitoring and Evaluation Assistants for Secondary Health

Android-based smartphones have been provided to districts supervisory officers (i.e. EDO, DDHO, and DHO) who have been tasked with visiting health facilities (BHU, THQ, RHC and DHQ) in their area. Supervisory Officers report data through Health Watch application for indicators like non-functional equipment, stock-out medicines and absenteeism, the supervisors take a self-photo that is submitted along with the data to authenticate that the officer visited the facility in-person. The reports submitted by these officers automatically pop up on a map in real-time with GPS accuracy of 5 meters and available for use by managers at various levels. These applications effectively help in solving 3 key issues (PITB Website)

- Compliance of supervisory officers
- Validity of data collected
- Timeliness of data reporting

Under this e-monitoring initiative, 210 android phones and SIMs with internet facilities were provided to the district health managers (DHMs). These DHMs follow pre-assigned monthly targets for visiting healthcare facilities in their respective jurisdictions. The DHMs submit their inspection data through the Health Watch application developed by the PITB. The data submitted through the application automatically pops-up on a map using GPS in real-time. Multiple other reports are available on a web-based dashboard for departmental review and decision-making. The Health Watch program has delivered exceptional results since 2015 to 2017.

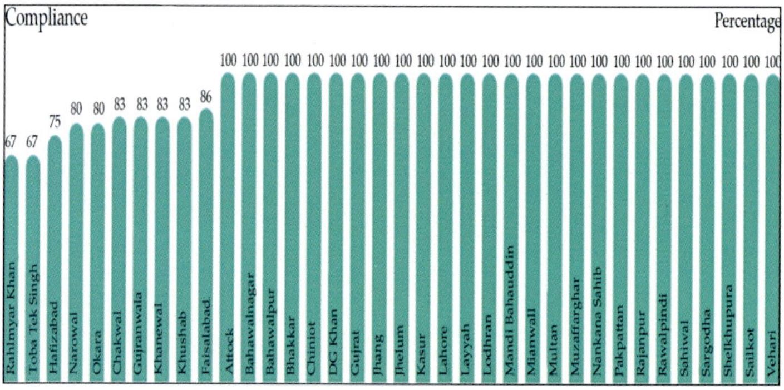

- The dashboard has received more than 110,000(https://www.pitb.gov.pk) entries from the field managers
- Staff attendance has improved from 22% to over 90 https://www.pitb.gov.pk)within a year (PITB Website)

Figure 3: Compliance

Pakistan has been facing the challenge of polio eradication since a long time. Smart applications to target the areas with high probability of cases and vaccination coverage have brought to the significant impact in assuring compliance of service.

"Health Line"

In 2010, Punjab was hit by the worst ever epidemic of dengue, affecting over 350,000(www.pitb.gov.pk) people. Due to the unprecedented scale of epidemic, the basic health infrastructure was not ready to address the rapidly changing requirements.

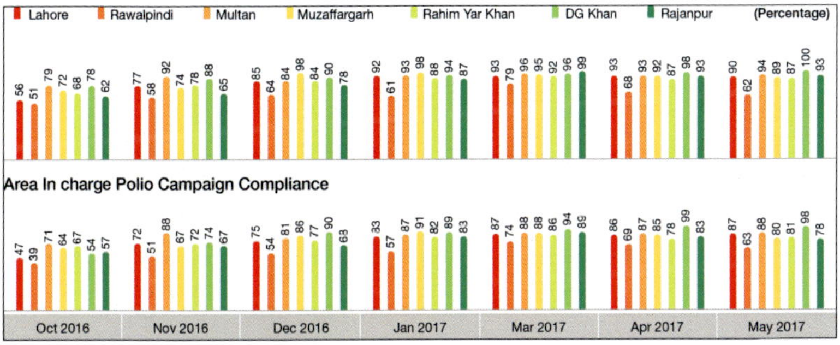

Figure 4: UCMO Polio Campaign Usage

PITB Website

The situation was aggravated as people did not even have basic knowledge of this disease. Thousands of citizens started contacting the relevant departments for information and complaints in panic. PITB rose to the challenge and took immediate measures to come up with a much needed solution. A Call Center was established in a record time of 48 hours for counseling the citizens 24/7 and was manned by 150 doctors taking calls on a toll free helpline.

The basic objectives included

- Patient counselling
- Awareness of preventive measures
- Advise on treatment protocol to health practitioners
- Logging of Fumigation Service requests
- Solid Waste Disposal requests
- WASA/ Sanitation related requests
- Complaints (Over charging by laboratories, Private hospitals, Govt. hospitals, etc.)

Disease Surveillance System

Health Science all over the world led administration with IT centric solutions which is manifested by the Centre of Disease Control and Prevention in United States health. The department use SatScan (software) and Early Aberration Reporting System (EARS) for preparedness to emergencies and to gather information to facilitate all stakeholders. PITB in 2011, developed a disease surveillance system, when Lahore witnessed a vital outbreak of Dengue fever and in response to geographical location which is effected by water borne diseases of typhoid fever, cholera, leptospirosis and Hepatitis A.

Figure 5: Dengue Tracking Application

PITB Website

According to the PITB non availability of historical data in an organized manner represents challenge for early detection of any future outbreaks thus limiting the capability of government for early preparedness and emergency response. To start with, Teaching Hospitals, Tehsil Headquarters (THQs) and District Headquarters (DHQs) are manned with dedicated data entry operators, responsible to report cases as per the predefined templates. Each data entry operator is equipped with laptop and internet dongle or any other means of internet connectivity. The data entry interface is exposed to government hospitals via secured usernames and passwords. Moreover disease wheels are provided to Basic Health Units(BHUs) and Rural Health Centers (RHCs) across Punjab which are used to report disease cases via SMS. These all consolidated information is used to make informed decisions making and to plan interventions to introduce innovative ideas to combat with the situation.

CONCLUSION

IT centric transformation of the management system brought about tremendous benefits for aligning departments in a way to produce maximum output while mobilizing the available resources. Pakistan following the same fashion have revitalized the health system by deploying IT trained personnel for strengthening the monitoring and evaluation mechanism with smart applications and sophisticated devices. IT integrated models will ensure transparency and accountability while generating the statistical data to form decisions on evidence. Fostering innovations for coherent policy design is the global demand which the governance in Pakistan has successfully implemented. However, some implications have its due share for non-compliancy of systems. The government can establish a balanced framework for incorporating holistic, comprehensive and coherent system of administration for ensuring human well-being, economic development and a healthy environment.

Recommendations and Observations:

- National ownership of health can design framework to establish credibility for international funding and public financing to allocate more resources for medical training, monitoring devices, capacity building, and technical resources.
- A credible liaison of local government and community can foster customer satisfaction by soliciting their needs and developing mechanisms to fulfil their demands.
- SDGs are achievable with the benefits of devolution, political will, less influence of bureaucracy while diminishing rural-urban and gender disparities.
- Policymakers in a mode of public-private partnership can create an enabling environment to integrate ICT models for strengthening accountability and ensuring transparency for harnessing informed decision and evidence based policy making.

REFERENCES

Book

1. Sarfraz Khawaja and Jahanzeb Waheed(2016), *IT Key to Good Governance, Governance in South Asia, ed.* Rumki Basu, M. Shamsur Rahman, Taylor & Francis (Delhi). p 84.
2. **Reports**
3. Economic Survey of Pakistan (2015-2017), *Ministry of Finance, Government of Pakistan* (Islamabad, Official Printers, 2017), 170.
4. Pakistan 2025 One Nation- One Vision development Programme, *Ministry of Planning, Development & Reform* (Islamabad, Official Printers, 2014).
5. Report of Inter-Agency regional analysis Network a Think-Tank, "Stunting in Pakistan: A Trend Analysis of Underlying Factors by 2030" (Paris, official printers, 2017), 1.

Websites

6. https://www.pitb.gov.pk
7. http://cerp.org.pk/
8. http://www.pide.org.pk/pdf/ASER-Pakistan.pdf
9. http://aserpakistan.org/index.php
10. uploads/2017/03/ENGLISH_Why_it_Matters_Goal_3_Health.pdf(accessed 4 December, 2017)
11. http://www.sdgindex.org/assets/files/sdg_index_and_dashboards_compact.pdf(accessed 8.12.2017).
12. http://www.bbc.com/news/world-us-canada-36675260(accessed 5.12.2017)
13. http://www.un.org/sustainabledevelopment/wp- / uploads/2017/03/ENGLISH_Why_it_Matters_Goal_3_Health.pdf(accessed 4 December, 2017)

Newspapers

14. Waqas A Khan, "ICT in education, reached? Nowhere!" *Pakistan Today* (June 18, 2017)
15. Zulfiqar A Bhutta, "*Malnutrition and stunting in Pakistan*" Daily Dawn (10 Feb, 2017).
16. Ikram Junaidi, "132,000 people in Pakistan living with HIV, survey finds" *Daily Dawn* (October, 01 2017)

Chapter 8

Waste Management as a Key Component of Sustainable Environment

Meeta Sharma

ABSTRACT

Sustainable waste management is a goal that all societies must strive to achieve. It is a main concern especially in the developing countries. Municipal authorities generally fail in providing adequate services due to insufficient funds and inefficient legislation in this regard. The increasing concern regarding the environment has led to the development of innovative practices of waste management. This paper deals with sustainable waste management, with a focus on the need to follow an expanded waste management hierarchy, to extract value specifically from municipal solid waste. It highlights the need for improvement in waste management at technological, institutional and financial level as there is a limit to nature's sustainability and waste management is a key ingredient to sustainable development. Emission of methane and carbon dioxide from waste could be responsible for increment in green house gases. Improper waste management has its own adverse health and environment impact. Focus should be shifted from efficient waste removal to waste minimization in order to be in tune with the goal of sustainable development. Rag pickers too can be trained and given logistic support for decentralized waste management. Community perception of fees of door to door waste collection should be gathered so as to increase community participation.

Keywords: Sustainable waste management, Waste management hierarchy, Municipal solid waste, Waste minimization, Community participation.

Nature has supported and maintained life on earth since time immemorial. This capacity of earth is known as sustainability of nature. People across the world have realized the detrimental impact of several human activities on nature and have came together to deliberate and adopt policy of sustainable development.

In 1980 the International Union of Conservation of Nature published a world conservation strategy that included one of the first references to sustainable development as a global priority and introduced the term "Sustainable Development". In 1987 the United Nations World Commission on Environment and Development released the report Our Common Future, commonly called the Brundtland Report. The Report included what is now one of the most widely recognized definitions of sustainable development "Sustainable Development is development that meets the needs of the present without compromising the ability of future generations to meet their own needs." (Zhang, Mi, Shen, January 2019)

Sustainability means taking into account all the resources available on planet earth such as humans, animals, plants, water, air, land etc and managing them to grow and develop so that the resources that exist at present may still be there for the future generations as well.

High rate of population growth, declining opportunities in rural areas and profitable occupations in urban areas largely contribute to urbanization. The cities have grown haphazardly in turn increasing pressure on urban governments. The increasing population has resulted in sanitation issues, congestion, environmental degradation, inadequate water and electricity supply etc. It poses a challenge to urban planners and citizens alike.

Environmental issues have traditionally been given low importance resulting in substantial damage to human health and reduced productivity. Cities are considered growth engines but the impacts of often unplanned communities have disastrous consequences for environment. Air, water and soil degradation occurs. Balancing developmental needs with the limited natural resources is a key parameter in the struggle for survival. We have to understand our basic requirements and dependency on resources that would be determinant of our very existence.

Municipal Solid Waste (MSW) includes what is thrown out by households and commercial sector. MSW is one of the harder types waste to manage since it has so many different components. In India, the makeup of waste is roughly 50% biodegradable, 25% inerts (construction and demolition waste, 9% plastic 8%paper, 4% rags and 1% glass.) (Zhu, Asnani, Zurbrugg, Anapolsky, Mane,

Nov. 2017) The composition of waste varies from season to season. In summers there is more biodegradable waste because of more vegetation the amount of plastic in waste has supposedly been decreasing due to the recent ban on plastic carrying bags in Rajasthan from August 2010.

IMPORTANCE OF WASTE MANAGEMENT

At present the world population is approximately 7 billion, and it is generating waste every day. But almost half of this waste isn't collected, treated or safely disposed, causing global waste crises. In percentage of total MSW generation, recycling (including composting) did not exceed 15 percent until 1990. The recycling rate grew more slowly over the last few years. The 2015 recycling rate was 34.7 percent.

Waste materials can either be solid, liquid, gaseous or radioactive. Waste management includes the collection, transport, processing, disposal, managing and monitoring of waste materials. It is required to reduce ill effects of waste on the well being and health of all living creatures and on environment. However waste management practices vary in developed and developing nations, and for residential and industrial producers. The goals of sustainable development cannot be met unless waste management is addressed as a priority. One of the major problems related to global environmental degradation has been the immense use of plastic in domestic and commercial activities as evidenced by many research reports, and a documentary produced by Public broadcasting system (PBS, USA, November 27/2019)

Table 1: Recycling as a Percentage of Generation

	1970	1980	1990	2000	2014	2015
Paper and paperboard	15%	21%	28%	43%	65%	67%
Glass	1%	5%	20%	23%	26%	26%
Metals	4%	8%	24%	35%	35%	34%
Plastics	Neg.	<1%	2%	6%	10%	9%
Yard trimmings	Neg.	Neg.	12%	52%	61%	61%
Selecte Conumer Electronics				10%	42%	40%
Lead-acid batter ies	76%	70%	97%	93%	99%	99%
Neg. = less than 5,000 tons or 0.05 percent						

https://www.epa.gov/facts-and-figures-about-materials-waste-and-recycling/national-overview-facts-and-figures-materials

The proliferation of plastics has been devastating for the planet earth and humans and animals residing on it. It is unlikely that there is a single beach in the world that doesn't carry plastic wrappers, straws, cigarette butts etc. Marine animals die when consume plastic products. The sheer volume of plastic bottles, grocery bags packaging is overwhelming. Most of the waste is produced by developed countries. Serious measures are being debated around the world to reduce it.

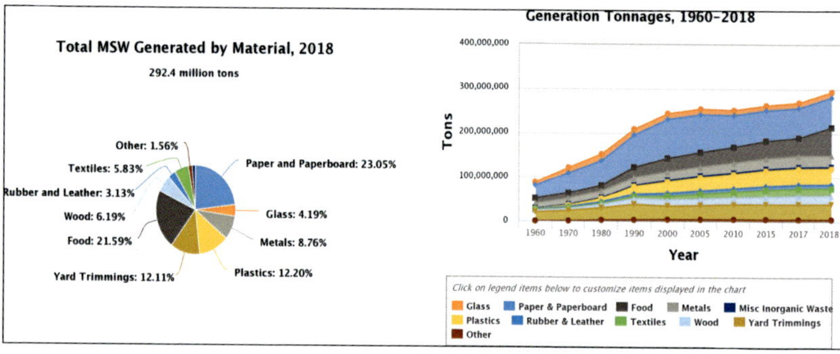

Figure 1: Total MSW Generated by Materials, 2015

https://www.epa.gov/facts-and-figures-about-materials-waste-and-recycling/national-overview-facts-and-figures-materials

Hazardous (medical waste) in large quantities are dumped in rivers and oceans by ignoring internationally accepted regulations against dumping. Multinational corporations have been found guilty of those malpractices. Rotting unmanaged waste produces air pollution.

E-waste components contains toxic and non-biodegradable waste which present health and environmental threats such as toxic smoke from recycling and leaching from dumping of e-waste into landfill sites into local water tables.

Mostly in poorer countries of the global south an efficient waste management is very low. Financial resources are scarce hence the waste is either openly burnt or simply dumped to rot creating green house gases. Waste is a key vector of disease and provides breeding grounds for mosquitoes.

HEALTH AND ENVIRONMENTAL IMPACTS

Some of the adverse environmental impacts of unscientific handling and indiscriminate dumping of solid waste are:-

- Ground water contamination by the leachate generated by the waste dumps.

- Surface water contaminations by the runoff from the waste dumps.
- Foul odor, pests, rodents and windblown liter in and around the waste dumps.
- Generation of inflammable gas (methane) within the waste dumps.
- Release of green house gases such as carbon dioxide.
- Epidemics through stray animals and other disease vectors.
- Inadequate waste management is the cause of serious urban pollution and health hazard.

The conventional approach to SWM has been to manage the removal of solid waste from the vicinity of human settlements resulting in the mechanized systems of collection and transportation of waste and landfills to bury the waste. It is realized that societies will not be able to deal with the growing responsibility of tackling the waste. The waste management had to change its focus from efficient removal to waste minimization in order to be in tune with the goal of sustainable development.

Municipal solid waste management has emerged as a big challenge because of health and environmental concerns, and also due to huge quantities of waste generated. Management of hazardous and non hazardous residential and institutional waste in the metropolitan areas is usually the responsibility of local government authorities. More than ¾ of solid waste management budget is allocated to collection and transportation, leaving scant funds for resource recovery, processing and disposal. Many waste utilization plants in India do not operate to their full potential, due to financial constraints and the absence of appropriate infrastructure, waste management therefore is neglected as a step child.

Twelfth schedule of Indian constitution makes it obligatory for the Municipal authorities to keep cities and towns clean. To implement these constitutionally mandated policies two ministries of Government of India namely 'Ministry of Environment, Forest and Climate Change' and 'Ministry of Urban Development' have initiated several policies and programs to improve current scenario of Municipal Solid Waste Management in India. Several laws have been enacted in this context.

The Environment Protection Act, 1986 provides an overall legal umbrella for regulating the process of disposal, and dealing with generated waste. Due to rapid urbanization massive waste management challenges are faced by municipalities.

Bio Medical Waste Rules (Management and Handling) Act 1998 prescribes that there should be a common biomedical waste treatment facility after every 150 kms for sanitary landfills and incineration.

Environment Ministry has also promulgated the Municipal Solid Waste Management Rules in 2000, which has now being revamped as Solid Waste Management Rules 2016.

While the Ministry of Urban Development has prepared a draft manual on Municipal Solid Waste Management to support cities and towns on planning and implementing a proper Municipal Solid Waste Management system in line with the Solid Waste Management Rules being promulgated in 2016. It has been observed that many policies and programs have failed to achieve their objective due to lack of clarity, awareness and poor enforcement by the regulators. (Mani and Singh, 2016)

In an effort to raise public awareness about the importance of waste management Swacch Bharat Abhiyan (Clean India) campaign was launched on 2 October 2014 at Rajghat, New Delhi by Prime Minister Narendra Modi. It aimed to achieve open defecation free India by 2 October 2019, the 150th anniversary of Mahatma Gandhi by constructing 12 million toilets in rural India at the projected cost of 1.96 lakh crore. This national campaign spanned over 4,041 cities and towns. It aimed to clean up streets, roads of cities, smaller towns and rural areas.

Ministry of Housing and Urban Affairs conducted its first survey 'Swachh Survekshan-2016' survey for ranking 73 cities (Urban Local Bodies), in January 2016, under the ambit of the Swachh Bharat Mission (Urban). In order to expand the coverage of cities, the Ministry conducted its second survey 'Swachh Survekshan-2017' in January-February 2017, to rank 434 cities. 'Swachh Survekshan 2018' saw a massive increase in both scale of survey and intensity of participation, with 4,203 cities covered in last year's survey. Swachh Survekshan 2019 has scaled even greater heights, covering 4,237 cities, and that too, in a record time of 28 days, in a completely paperless, digital format for data collection! Moreover, even in this short span of time, assessors managed to visit nearly 73,000 wards, 21,000 commercial areas, 69,000 residential areas, 75,000 community/public toilets.

The Minister informed that we have made impressive achievements under the SBM in these four and a half years wherein urban areas of 23 states / UTs have become ODF, and more than 94% cities are already ODF. Nearly 63 lakh individual household toilets (94% progress), and more than 5 lakh community / public toilet seats (more than 100% progress) have been constructed. Parallel to it more than 42,000 public toilet blocks across 1400 cities have been mapped and visible on Google maps. The Google toilet locator also provides an option for citizens to provide their feedback after using the toilets. Waste processing has gone up to 52% compared to a mere 18% at the start of the Mission. (http://pib.gov.in/pressreleaseshare.aspx?PRID=1567704)

WASTE MANAGEMENT HIERARCHY

At present most of the recyclable waste ends up in dump yards, due to lack of efficient waste management. It is estimated that urban India alone generates about 42 million tonnes of municipal solid waste per annum. Since by 2050, 50% of country's population is projected to be urban and the amount of waste will also increase substantially. (Gupta and Gupta, May 2015) Municipal and private establishments must cooperate to face this challenge.

Waste segregation can provide employment to many rag pickers. In some urban areas people working in informal sector collect waste from each household to get collection fee and have additional income from the sale of recyclables. It ensures that less waste reaches landfills. Greater community participation to encourage such efforts could be very helpful. It should be noted here that waste should be segregated at the source and go through different methods of processing. Sanitary landfills are the ultimate places for disposal of unutilized waste despite the cost involved in transporting it.

It has been observed that waste management personnel should work not only on waste management system but also on production system. Waste sector needs to work in synergy with the production and manufacturing section to promote the production of goods and material with minimal hazardous contents resulting in least amount of waste.

There is no subject in the world more vital to the future and sustainability of the planet earth for future generations than that of waste management and all that it encompasses. Animals produce organic waste only. Human beings in their ignorance have now created so much inorganic waste that the whole planet is suffering from pollution in air, in rivers and oceans of the world as well as on the land masses.

The different waste management options can be organized in an order known as 'waste management hierarchy' that reflects relative sustainability of each.

- Minimization (most favoured)
- Reuse
- Recycle
- Energy recovery
- Disposal (least favoured)

One of key principles underlying sustainable waste management is to ensure that waste is dealt with as high up waste hierarchy as possible. **(United States Environment Protection Agency, May 2018)**

Since every waste disposal option has some impact on the environment, therefore the only way to avoid impact is minimization of waste production. Therefore it has top place in hierarchy. Reuse is followed by recovery techniques (recycling, composting and generating energy from waste). Disposal to landfill or by incineration should be the last resort. Environmental and economic costs and benefits of disposal channels must be part of policy making and serious political discourse. Transporting waste has economic costs therefore it should find place near to the waste generation sites.

Rubbish is made of different materials and should be treated differently – some should be reused, some should be recycled or composted, some should be burnt and other buried.

COMMUNITY PARTICIPATION

Environmental degradation due to excessive waste generation and incapability to tackle the waste, impacts the urban poor who do not have fair access to sanitary services in the city and are subject to extremely unhygienic conditions.

Driven by compulsion to earn daily bread the poorest get involved in waste picking and sell it to the scrap dealers and recycling industry. Organized groups of rag pickers can be trained and given logistic support for decentralized waste management. They should also be provided health insurance.

Waste minimization and community based waste management seems to be the only sustainable way to manage the waste.

To achieve the desired success continuous participation of the community is required for example to store the garbage in specific bag or bin, separate dry and wet waste etc. But often participation of households in these activities is quite low, low willingness to pay for waste collection services and lack of interest in keeping public places clean is observed.

To gain maximum contribution of community proper consultation with community on the desired service, appropriate incentives for the households who contribute can be suggested.

Households often behave contrary to the schedules and rules of effective solid waste management due to lack of facilities. If transfer point or dumping site is more than 100 meters away from the household then the people find it easier to throw their waste much more often in open spaces.

Often municipalities lack the money or the manpower to fulfill this task, it is difficult to seek cooperation from low income households because of their inability to pay the fee levied by municipalities for waste collection. Therefore it is only by making these poor people realize that despite the fee it is in the long range interest of individuals and the community well being. A solution may be to charge fee according to individual's capacity.

The dream of cleanliness can be achieved if 125 crore people of this country join hands. Jan Bhagidari or public participation is required to achieve success in this regard. Swaraj was achieved by satyagrahis, in the same way shrestha bharat would be achieved by swacchagrahis (those who work for cleanliness) Swacchta needs a revolution of ideas.

REFERENCES

1. Environment and urbanization, Solid waste management and sustainable cities in India: the case of Chandigarh, journals.sagepub.com, 24.01.18
2. Sustainable waste management, https://torinobyveg.worldpress.com, 24.01.18.
3. Sustainable Development, https:en.m.wikipedia.org>wiki>, 28.01.18.
4. Zhu, Da Asnani, PU. Zurbrugg, Christian (Dec.07) Improving Municipal solid waste management in India: A source book for policy makers and practitioners. Herndon, VA, USA: World Bank Publications.
5. http://www.indiaenvironmentportal.org.in/reports-documents/, 5.11.19
6. https://pib.gov.in/newsite/PrintRelease.aspx?relid=189231, 4.11.19
7. https://www.teriin.org/sites/default/files/files/factsheet.pdf
8. https://www.unenvironment.org/annualreport/2018/index.php#cover
9. https://wedocs.unep.org/bitstream/handle/20.500.11822/25496/singleUsePlastic_sustainability.pdf?isAllowed=y&sequence=1
10. http://datatopics.worldbank.org/what-a-waste/tackling_increasing_plastic_waste.html

Chapter 9

UJAMAA: An Alternative Model of Development

Radha Kumari

ABSTRACT

The debate over crisis and limitations of the present model of development has forced scholars to mull over the alternative ways of development. With the growing popularity of post –colonial discourse in academics, attention has now been shifted to study tribal life-styles and their models of society specially indigenous models of African tribes. Tanzania is an African country where 'Ujamaa model' was applied by Mr. Julius Nyerere, who was a political leader and the president of Tanzania. According to him, socialism and democracy are the ethos of this model. Nyerere's claim of this model being indigenous implies that this is rather a unique model in its characteristics, not a copy of any others. In this paper an attempt is made to critically evaluate the success and failures of the model.

Keywords: Ujamaa, Socialism, Democracy, Development, Indigenous Model.

The entire discourse around development in nineteenth and twentieth century has been largely euro-centric in which the 'west' represented 'Development, Civilisation and Culture'. Whether it was economy, politics or society, West was seen at the apex of the uni-linear path suggested by western scholars like

W.W. Rostow and East was expected to follow it blindfolded. (Rostow, 1960, 4-16) With the end of the cold war in 1991, Francis Fukuyama endorsed this thesis by writing his classic 'The End of History' which declares that the zenith of ideological evolution has already been reached and capitalist model of development within the western liberal democratic framework has emerged as the best possible model (Fukuyama,1989, 3-18).

But a series of crises at the dawn of twenty first century has shown that euphoria around this victory was short-lived.

The World, since then has witnessed Iraq war, economic recession of 2008-09, Greek financial crisis, Syrian civil war (a major cause for horrific numbers of migratory population) the ugliest and the cruellest form of terrorism and many more such problems which actually make it obligatory to question the status of development framework of West as an unquestionable destination. The worst of all, if one may say, is the threat for survival of human kind due to continuous ecological degradation, a consequence of the same development model which the 'West' has been boasting of.

HISTORY OF TANZANIA

Before the world War I, most of Africa was being colonised and exploited by European powers and were divided into many zones for the purpose. Tanzania was under the dominance of Germany and was part of 'German East Africa'. After the end of the World War I, administration of this land came under the British mandate provisions with a new name 'Tanganyika'. Though, Britain encouraged the indigenous administrative institutions, yet these institutions were not representative of native Africans as African members were not elected to them.

After the end of the world war II, Tanganyika was kept under trusteeship administration untill 1947. During the decade of independence of Africa, Tanzania also became independent and it adopted its republic constitution in 1962. Julius Nyerere was elected as president of Tanzania in coming elections.

In 1964 Nyerere reached an agreement with **Abeid Karume**, the president of the offshore island of Zanzibar which has been so closely linked in its history to the mainland territory of Tanganyika. The two presidents signed an act of union, bringing their nations together as the United Republic of Tanzania. Nyerere became the President of the new state, with Karume as his Vice-President. Nyerere declared his political creed in a document of 1967 known as the Arusha Declaration. The declaration introduced a socialist state and was followed by the nationalization of key sectors of the economy. (http://www.historyworld.net 2016))

PATH OF DEVELOPMENT FOR TANZANIA

With the onset of era of independence, a new form of dependence started in African countries. Competition for resources and influence zones emerged between the two super-powers engaged in a new tug of war in the name of ideologies. Newly independent states of third world were supposed to take one of these sides and join their respective alliances. Julius Nyerre was never interested in capitalistic mode of development as he had a commitment to equality. According to Nyerere, western form of liberalism had attached more significance to liberty with less concern for equality. In a capitalist system, Nyerere opines, affluence is attained by exploitation of men by men which leads to a polarised society along the lines of classes. Whereas, Nyerere's prime focus was on self reliance which according to him, would be endangered by giving in to western values and patterns of organisation. In his words, "Given the present inequalities between nations, capitalist development is incompatible with full national independence for Third World countries. For such development will be foreign owned and controlled; local capitalists will be subsidiary and will remain subsidiary. [Capitalism is also] basically incompatible with....human dignity and self respect for all, with equal freedom for all inhabitants of the society. For capitalism means that the masses will work and a few people-who may not labour at all- will benefit from the work. The few will sit down to a banquet and the masses will eat whatever is left over. [Thus capitalism is] incompatible with national purposes of all Third World governments."[1] (Fattom, R., 1985, pp-3-24)According to Nyerere the western notion of socialism was not a suitable path for Africa. Though, Nyerere supports socialism as he is a votary for equality; yet he did not want to trade Democracy for socialism. He was not in favour of kind of revolution which ....

He believed that European socialism was based on the premises of class structures and the inevitability of conflicts among these classes. In his words:

"....European socialism was born of [the] conflict [between the forces of agrarian and industrial revolutions and] its apostles sanctified the conflict itself into a philosophy. Civil war was no longer looked upon as something evil or something unfortunate, but as something good and necessary....Brought up in tribal socialism, I must say I find this contradiction quite intolerable. It gives capitalism a philosophical status which capitalism neither claims nor deserves for it virtually says 'without capitalism, and the conflict which capitalism creates within society, there can be no socialism! This glorification of capitalism by the doctrinaire European socialists, I repeat, I find intolerable. African socialism, on the other hand, did not have the benefit of the agrarian revolution or the industrial revolution. It did not start from the existence of conflicting 'classes' in society. (Fattom, 1985 3-24)

NYERERE'S IDEA OF SOCIALISM

Socialism, as an ideology can be called a broad church where differences of opinion exist to varying degrees. The only unifying theme in different variants of socialism should be equality. Nyerere was a votary of equality in society and to achieve that he wanted to establish a political system built on the ideology of Socialism. As has been discussed earlier, he did not want to import and imitate the idea from the 'West'; he offered a model of socialism which, according to him was indigenous, rooted in African Society.

As Nyerere conceded that he had a tribal upbringing, therefore he never claimed of having innovated any original and unique form of socialism which did not exist before. Rather he had a firm belief that African society used to live with the practices of indigenous form of communalism which unlike European form of socialism was based on cooperation. Julius Nyerere admitted that the African society and economy was in need of modernisation and restructuring because the colonial government corrupted the ethos of African socialist essence. Nyerere wanted to rebuild Tanzanian society on its traditional and indigenous roots which would help in achieving more humane form of development.

Nyerere preferred African way of development for other reasons also. In his views, heavy industrialisation which was the feature of both the models of development (whether capitalism or socialism) required a large amount of capital which could be arranged only by taking loans from rich countries. According to Nyerere, this was the cause of third world countries not just being trapped into cycle of debts but also of loss of control of these countries' governments on their own economies as well as on their society. In Arusha Declaration, he clearly elaborated on this subject: "Independence means self reliance. Independence cannot be real if a nation depends upon gifts and loans from another for its development.....Gifts [however] which increase, or act as a catalyst to our own efforts are valuable"[2] (Nyerere, 1968, 239-240))

Analysing the views of Nyerere, Fatton observed, development was not possible without the adoption of a strategy of self reliance. Self –reliance entailed the conviction that the success of African socialism depended upon Africa's capacity to rely on its own resources, energies and culture for its autonomous development. This implied neither chauvinism nor autarchy; the concept of self-reliance forcefully opposed dependence on external sources of finance, capital and technology because such dependency allowed alien forces to make vital national decisions and subsequently control the development process. Reliance on foreign assistance would never provide the necessary amount of resources required for developing a poor country like Tanzania. And even if such assistance would suffice, it would endanger the policies of African socialism by entrusting them to alien powers." (Fattom, 1985, 3-24)

AFRICAN MODEL OF SOCIALISM

The main language of Tanzania is 'Swahili'. Ujamaa is a Swahili word which means familyhood. According to Nyerere in traditional African society, society used to care for all members whether a person was rich or poor, so it was a kind of an extended family in which despite natural calamities or any other problems, no one would ever starve of food and human dignity. If a person lacked personal wealth he could always rely on the wealth of society as a member of it. This was African socialism. So in African socicties members were free of all kind of insecurities which are generally responsible for acquisitive tendencies among its members. According to Nyerere, "Acquisitiveness for the purpose of gaining power and prestige is unsocialist. In an acquisitive society wealth tends to corrupt those who possess it. It tends to breed in them a desire to live more comfortably than their fellows, to dress better, and in every way to outdo them they begin to feel they must climb as far above their neighbours as they can...... apart from the anti-social effects of accumulation of personal wealth, the very desire to accumulate it must be interpreted as vote of 'no-confidence' in the social system. (Nyerere, 1962, 66)

Nyerere pointed out many more features of African socialism. In his words, "In traditional African society everybody was a worker. There was no other way of earning a living for the community. Even the elder who appeared to be enjoying himself without doing any work and for whom everybody appeared to be working, had in fact worked hard, all his younger days.......it was taken for granted that every member of society- barring only the children and the infirm- contributed his fair share of efforts towards the production of its wealth." (Nyerere, 1962, 67-68)

Nyerere further elaborated the concept of 'worker' in traditional African society. It was not synonym of employee juxtaposed to 'Employer'. Rather, In African society 'worker' meant a person engaged in labour for one's livelihood with human dignity. In African society employer term was alien which was introduced by colonial people who had a capitalist attitude of mind, using other's labour to accumulate wealth. According to Nyerere this attitude was inculcated among African people also. According to Nyerere this attitude was incompatible with the socialist society. Nyerere wanted to change this attitude of using others and exploiting others for accumulation of power and prestige.

Though, Nyerere took too much pride in boasting of the tradition of hospitality in African society, yet, he was also aware of the saying: " MGENI SIKU MBILI; SIKU YA TATUM PE JEMBE" (Treat your guest as a guest for two days; on the third day give him a hoe!) In fact, the guest was himself supposed to ask for the work as staying without work for a longer period was considered shameful. (Nyerere, 1962, 68)

Nyerere further elaborated the concept of 'worker' in traditional African society. It was not synonym of employee juxtaposed to 'Employer'. Rather, In African society 'worker' meant a person engaged in labour for one's livelihood with human dignity. In African society employer term was alien which was introduced by colonial people who had a capitalist attitude of mind, using other's labour to accumulate wealth. According to Nyerere this attitude was inculcated among African people also. According to Nyerere this attitude was incompatible with the socialist society. Nyerere wanted to change this attitude of using others and exploiting others for accumulation of power and prestige.

Nyerere also talked about the land as the most important tool of production which according to him was made a marketable commodity by foreigners. Nyerere said that in traditional African society each individual within society had a right to use the land as it was associated with the right to life. No one could have earned his livelihood without using some means of maintaining life so right to land was simply right to use it but the concept of landlords which recognises individual ownership of the land was totally absent in African tradition. According to Nyerere such landlords who earned their livelihood by charging rent for their land or possessed the land just for appreciation of its value by the surrounding development were also looters.

In a new system, Nyerere wanted that a member be entitled to a piece of land on the condition of using it. He wanted, Unconditional or 'freehold' ownership of land (which leads to speculation and parasitism) to be abolished by using legislation in harmony with socialist principles. In Nyerere's words those who eat without labour like idlers, loiters are parasites. One can hear echo of Gandhi's emphasis on physical labour for earning the livelihood. According to Nyerere, "there is no such thing as socialism without work. A society which fails to give its individuals the means to work, prevents them from getting a fair share of the products of their own sweat and toil, needs putting right. Similarly an individual, who can work- and is provided by society with the means to work- but does not do so, is equally wrong. He has no right to expect anything from society because he contributes nothing to society."(Nyerere, 1962,-68)

So 'Ujamaa' which means 'familyhood' is the foundation of African socialism. Nyerere believed that unity which African people showed during their anti colonial struggle would help them get back the socialist attitude of mind of tribal days, (the attitude which finds individual as embedded in a society, which is far from the temptations of personal gains, according to which an individual was poor or rich when the whole was poor or rich) and this would not only be limited to their tribe, but will spread throughout the nation even to the whole continent of Africa.

Fatton summarises 'Ujamma'; "Thus Ujamma is grounded in unique humanity of Africa, it is a product of the classless pre-colonial African heritage. In this sense, pre-colonial Africa's socio- economic arrangements were based on a communal form of socialism which articulated a sense of mutual respect between all the members of community who had the obligation to work and share their property and production. Mutual respect, the sharing of property and production and obligation to work embodied what Nyerere had called the principles of life of Ujamma. (Fattom, 1985, 3-24)

NYERERE'S ROADMAP TO UJAMMA

As noted above, Nyerere wanted to transform the African society into a socialist one but he was equally eager to have more participation at local levels; and for that purpose, tried to decentralise administrative structure. Nyerere was not a romantic philosopher, rather, he tried to experiment what he believed in and during his life time only. Louis A. Picard calls these experiments as unique. "Tanzania's leadership is trying to transform society into socialism while at the same time decentralising its administrative structures and its decision making process. In doing so, Tanzania hopes to avoid some of the dysfunctions of the over concentrated centrally planned economy so characteristics of many socialist political systems"[4] (Picard, 1980, 439)[5]

Soon after the independence of Tanzania, Nyerere realised that existing institutions of democracy were not reflecting the sentiments of Tanzanian people. Its foreign policy was also clearly inclined towards western countries or the capitalist bloc led by United States. He didn't take much time in his assessment that these countries were not genuine in their intentions; instead, using newly emerged third world countries as their puppets to fulfil their vested interests. Nyerere concluded that political system of Tanzania was not capable to rectify this. TANU (Tanzanian African National Union) was the only popular organisation because of its role in anti colonial struggle. Though elections took place in 1960, yet 58 out of 78 candidates were elected unopposed. In effect a majority of Tanzanian people could not be part of electoral process. Moreover, these candidates had developed their personal ambitions and were now eyeing only at on the national executive and assembly positions.

Nyerere acted swiftly and dramatically to correct this. Just forty-five days after independence on 23 January 1962, he resigned as Prime Minister. For eleven months he developed himself entirely, as president of TANU, to a revitalization of the party and to the mobilization of the people to fully engage in a unifying national endeavour to build a new and developing Tanganika[6] (Pratt, 1976, 116-117)

By 1965, Tanganyika adopted a hybrid constitution, which appeared self-contradictory in terms i.e. 'Democratic one- Party State'. Though the ideological foundation of this state was socialism, yet, it should never be confused with Leninist model of vanguard party. There were reasons for such conviction as membership of restructured 'TANU' was open to all and anyone could run for election for any representative body in the country. In each constituency, a large and representative body, the annual district conference of TANU, ranked the candidates for the election to the national assembly in order of preference. The national executive committee of the party then decided which two candidates would appear on the ballot....The elections which then followed, operated within a set of rules that were designed to ensure as fair a contest as possible. No candidate could have spent money on his own campaign. All election meetings in every constituency were organised by the party and were to be addressed by both candidates. No tribal language could be used at these meetings and no appeal for votes could be made on grounds of race, tribe or religion. No politician or other prominent Tanzanian could campaign on behalf of any candidate.[7] (Pratt, 1999, 142-143)

This was also true that such type of design brought significant transformation in the character of representative institutions of Tanzania. Nyerere's achievement in this regard was that he was able to replace those ones from politics, with whom people were not contented, that too in a peaceful manner and avoiding the divisive impact of cultural and ideological factors. Yet, Nyerere himself admits on one occasion that his 'Democratic state' was undemocratic on occasions where detentions and breaches of human rights were defended in the name of nation's stability. In 1967, Tanzanian government came up with 'Arusha Declaration' on policies of socialism and self reliance.

On the economic front, Nyerere was not only committed to equality of initial opportunities, but also, to the goal of people enjoying almost equal status as a member of society. He was not very happy about the people in bureaucracy and politics, developing acquisitive tendencies. He was even concerned about the wide gaps in salary levels of officials and wanted to rectify that. He realised very soon that he was not getting much support for his efforts. In agriculture his one of the main concerns was production of cash crops on single holdings which according to him would led to the stratification of countryside of Tanzania.

He took many steps for making Tanzania genuinely socialist according to his own vision. These steps included:-

- Nationalisation of banks
- Nationalisation of important parts of limited industrial sector
- Introduction of leadership code to contain corruption and to block the emergence of private sector economies

- A sustained effort to reform the educational system of Tanzania to produce socialism compliant youth.
- Movement of peasants from their rural holding to newly created villages
- An extensive takeover of household properties which were not occupied by their owners.
- Movement of capital from Dar Es Salaam, to Dodoma, a more 'African' town in the centre of the country.

One can easily conclude here that all these steps were initiated with a genuine determination for development of a socialist and prosperous Tanzania. Above all, his focus was on "the creation of a country in which the people took a full and active part in the fight against poverty, ignorance and disease"[8] (Pratt, C., 1999, pp-142-143)

UJAMAA IN PRACTICE: AN ANALYSIS

Nyerere, who was an idealistic socialist by instinct guided his country along lines which often had a utopian touch. Local self-sufficiency was emphasized. Traditional and simple solutions were sought for local problems rather than relying on technological foreign imports. Great importance was placed on education and literacy, in which excellent results were achieved.

With such policies Nyerere inevitably had to rely on external help -The Eastern Socialist bloc in general and China in particular. Nevertheless he was able to keep his non-alignment policy.

The Arusha Declaration put agriculture at the centre of the national economy and introduced a programme of 'villagization' - meaning the moving of peasant families into cooperative villages where they could supposedly work together more productively. As elsewhere where such cooperatives were tried (in particular Mao Tse-tung's China, a source of inspiration to Nyerere), they proved both unpopular and inefficient. When Nyerere relinquished executive power voluntarily in 1985 (a rare act in modern African history, and certainly one with no appeal to Mao), he admitted that his economic policies had failed.

In his twenty-three years in office he established an impressive reputation as an independent and free-thinking African statesman - willing to sever relations with the UK (1965-8) because of British support of racist Rhodesia and South Africa, (http://www.historyworld.net 2016)

As, the history shows, every decision has its unintended consequences also. Despite, showing Stateman's leadership and characteristics, Nyerere realised that most of his steps had proved to be counter-productive. Tanzania was still not free of poverty and stratification of people. Even the peasantry was not happy with his decision of rehabilitation of them from their original surroundings to Ujamma Villages. Industrial output also came down and country, instead of becoming a self -reliant one, became more dependent on foreign assistance. Amid all these developments, Nyerere, himself, could feel the nerve of the country and decided to step down in the mid of 1980s.This was a peaceful change within the constitutional framework of Tanzania from a single party state to competitive multi-party system.

Reasons for Failure of Ujamaa Model In Tanzania

In hindsight many reasons can be cited for failure of his model. Some even find the basic belief of Nyerere to be faulty regarding the communitarian attitude of Tanzanian people which was more conducive to the system of collective ownership and communal agricultural activities. They questioned that had it been so, how could have the Tanzanian people be individualistic by their attitude during colonial period. His view of not adopting heavy industrialisation was also seen as one of the reasons for backwardness of Tanzania. Other than the socialist measures taken by Nyerere's government, Tanzania also faced the severe drought these years. International events were also not favourable for Tanzania during the decade of 1970s as due to oil crisis; country had to pay a lion's share of its reserve for energy needs.

CONCLUSION

Earlier in the discussion, I referred to Fukuyama's thesis which claimed victory of western liberal model of capitalism. Can this be concluded here that Africa and Asia like poor continents are to follow this model only for development? I am reminded of Gandhian model of development which has some similarity with Nyerere's ideas. Like Nyerere, Gandhi was also against the blind imitation of both i.e. Capitalist and European Socialist models. Gandhi also talked of reviving India's old structure of small and cottage industries with his prime focus on self reliant villages. The importance of physical labour has been emphasised in ideas of both the leaders as a means to achieve a just society. Both had equal concern for ecological protection. Both the leaders supported peaceful revolution only and discouraged violent methods. Though, there were more differences than similarities between their views. As Gandhi never assumed any post in free India and he lived even less than a year after India's independence, it is very difficult to infer what fate would his model of development have met, if exercised.

But at the same time there is no dearth of such examples even in Africa where capitalist model was applied and the people of these countries faced more severe economic challenges than Tanzania. In my view failure of this model does not make it less worthy of scholars' attention who keep exploring alternative models of development. In a world, facing daunting challenges of environmental degradation, terrorism, growing racism, Nyerere could be called a social engineer who used the principles of dignity of labour, equality, environmental protection and self- reliance for reconstruction of his country. The most important and the least discussed aspect of his political leadership was his belief in peaceful change and democracy. When many of the African countries werc under the fire of civil war because of the elites who were deeply drenched in corruption, not coalescent, but divided, power hungry and had to be ousted by force only; Nyerere was a genuine leader who stepped down voluntarily.

REFERENCES

1. Fattom, R., (1985), The Political ideology of Julius Nyerere : The Structural Limitations of "African Socialism", *Studies in Comparative International Development,* Vol. 20, Issue.2, pp-3-24 https://link.springer.com/article/10.1007/BF02718489, accessed on 15.7.16
2. Fukuyama, F. (1989). *"The End of History?". The National Interest (16): 3–18. ISSN 0884-9382. JSTOR 24027184.*
3. Nyerere, Julius, K. (1968), *Freedom and* Socialism, Oxford University Press, London.
4. Nyerere, Julius, (1962), 'Ujamaa': The Basis of African Socialism in: Hord, Fred Lee (Mzee Lasana Okpara), Lee, Jonathan Scott, 'I am because We are: Readings in Black Philosophy', University of Massachusetts Press, Amherst, USA.
5. Nyerere, J., (1967), **Written:** for Tanganyika African National Union, **transcribed by:** Ayanda Madyibi. The Declaration was discussed and then published in Swahili. This revised English Translation clarifies ambiguities which existed in the translation originally issued. https://www.marxists.org/subject/africa/nyerere/1967/arusha-declaration.htm accessed on 15.7.16
6. http://www.historyworld.net/wrldhis/PlainTextHistories.asp?historyid=ad23#ixzz4demY5Mht accessed on 15.7.16
7. Picard, Louis, A.,(1980), Socialism and Field Administrator: Decentralisation in Tanzania, Comparative Politics, The City University of New York, Vol. 12, No. 4 (July), p-430-437
8. Pratt, Cranford, (1976), The Critical Phase in Tanzania, 1945-67: Nyerere and the Emergence of a Socialist Study, Cambridge University Press, Cambridge.
9. Pratt, Cranford, (1999), Julius Nyerere: Reflections on the Legacy of His socialism, *Canadian Journal of African Studies,* Vol.33, published by Canadian association of African Studies, pp-137-152, url: http://www.jstor.org/stable/486390 accessed on 27.9.11
10. Rostow, W. W. (1960). *The Stages of Economic Growth: A Non-Communist Manifesto*. Cambridge University Press.

Chapter 10

Global Governance and Sustainable Development: Emerging Realities and Perspectives

Dhirendra K. Vajpeyi and Sheila Rai

ABSTRACT

From ancient times — be it Plato's quest for an ideal polity, Kautilya's suggestions about state-craft, or Confucian's concern on morally based harmonious social order — to date philosophers, policymakers, and religious leaders have pondered and analyzed issues related to human destiny and the "fate of the Earth in the universe". (Xi Jinping, January 18, 2017). As societies become complex the search for viable answers to these fundamental core questions seem to elude us. Distribution of basic necessities — food, shelter, personal security and political stability — requires leaders and institutions to implement broad societal policies. In the absence of good governance humanity faces Hobbesean nightmare where life is nasty, brutish and short". The pivotal question as to what is good governance and how to achieve it without imposition from one system — values, politics, culture — over others has been debated, and often times it seems that there is no one simple answer/model.

Kewords: Globalisation, Global governance, Accountability, Indian / China Model.

According to Thomas Jefferson, good governance should take "care of human life and happiness, and not their destruction...is first and only legitimate object of good government". Good governance must be inclusive based on global consensus and not only on Washington consensus. It must serve "equally the rich and the poor" (Annan 2001:126). It cannot be realized by Western or for that matter non-Western values and models alone. It is unlikely to be achieved "by attempts to rebuild (so called) liberal order on Macron-style yuppie populism, inclusive nationalism, pragmatic patriotism (China) or any other expedient of an intellectually insolvent (though materially resourceful) centrism" (Mishra 2018:13). Global governance must entail "diplomacy and politics within and between alliances and coalitions of states, international pressure groups of non-governmental actors...feedback loops between international, national and sub-national actors; and institutional procedures and methods that channel inputs and determine their efficiency". (Finkelnstein 1995: 368-369).

This paper comprises two parts. Part I will address main themes of the anthology by analyzing and discussions on the following issues:

- Nature of global governance in general
- Globalization and global governance
- Impact of technology on global governance
- Global governance and sustainable development
- Multinational corporations, non-governmental organizations, and global governance
- Present structure of global governance institutions — the United Nations, the World Bank, the IMF, the WTO, etc. — The Genesis
- Discontent with present structures and the need for reforms and the future directions

The second part will analyze much discussed two alternative models of development and the rise of China as a global power and its role in shaping global governance (Global Governance with Chinese Characteristics), and the Indian model in a democratic open society.

DEFINITION AND NATURE OF GLOBAL GOVERNANCE

Let us start with defining global governance. It is a multi-dimensional and complex concept. Perceptions and definition of global governance have been influenced by political, economic, cultural, geographical, and social differences around the world. Several scholars maintain that the concept is too complicated to be operationalized. Defining global governance has been like asking ten blind people to describe an elephant, however, "in contemporary literature global governance is often used as an umbrella term that refers to rules, regulations, and decisions concluded by a whole network of international governmental institutions, NGOs and other global financial bodies." (Bakari 2018: 117). Chasek et. al (2010: 370) posit that "global governance in essence, can be described as the political interaction of States, international organizations and other transnational actors aimed at solving problems that affect more than one state or region." Finkelstein (195: 369) observed that global governance is "governing without sovereign authority, relationships that transcend national frontiers and doing internationally what governments do at home." Several other scholars have summarized six different elements of global governance (Jain 2005; Dwivedi 2013; Vajpeyi and Oberoi 2018; Koppell 2010):

- Accountability — includes transparency, liability, controllability, responsibility, and responsiveness
- Effectiveness — quality of policy making and distributive capacity
- Political stability
- Quality and degree of regulation
- Rule of law
- Control of corruption

It is also to be noted here that, "each global governance organization is part of an ecosystem populated by potential partners and political competitors — non-governmental organizations, private corporations, national and other domestic government agencies and regional transnational entities. The reality of competition and coordination in global governance ties together two issues of great interest to contemporary students of politics: regulatory competition and the rise of policy networks." (Koppell 2010:247) Another related issue is why of the global governance - the 'logic of global governance'. It is obvious that national states, NGOs and other organizations make a conscious and concerted effort to join and participate in such organizations — the UN, WTO, the World Bank, the IMF, etc. There is no coercion to make any organizations/ agency to join them. Supporters of Realist Theory maintain that these efforts are conditional by self-interest of the joiners. Participant members benefit from the actions of these global governance organizations. On the other hand the Institutionalists believe (and here both Realists and Institutionalists agree that the international system without rule based governance is anarchic) that global governance institutions not only provide opportunities and platforms to vent out differing views but also facilitate cooperation and commitment to 'collective goods' (Wendt 1999), build a system of commitment and enforcement that makes international political arena less chaotic and anarchic. "In a way this view reflects Marxist interpretations which see the capitalist demands of the international market as drivers of transnational cooperation rather than State interests" (Murphy 2002)

Table 1: Five Conceptions of Accountability

Concept of Accountability	*Key Determination*
1) Transparency	Revelation of facts, related to organizational performance
2) Liability	Organizational responsibility for its performance (consequences)
3) Controllability	Follow the organizational dictates
4) Responsibility	Follows the organizational rules, laws, professional standards
5) Responsiveness	Organizational ability to meet expectations of its clients

Source: Koppell (2010: 30)

GLOBALIZATION AND GLOBAL GOVERNANCE

A) Globalization

The end of WWII witnessed several majors changes in the world's political and economic structures — the end of Fascism in Europe, decolonization in Asia and Africa, the cold war between two ideological and political contenders, and creation of financial and political institutions such as the United Nations, the World Bank, and the IMF. The two superpowers — The USA and the USSR — competed to influence global politics by presenting two models of economic development. The US led Western model of Modernization/Globalization, championed a liberal approach. "It derived from the model of social order of free market economics...low degree of state intervention and reliance on indirect incentive to achieve social ends by assessing social problems in economic or quasi-economic (social indices) terms by participatory measures assuming full and fair access to the political market place, and by the "best public policy decisions to be derived from the balance of competing forces like one in the United States of America" (Reppy 1979). The Soviet (Marxist-Leninist) model on the other hand marketed the '*Social Solidarity*' model which emphasized individuals integration into the larger community with societal goals and ends mainly espoused by Saint Simonian managerial variety. It espoused more interventionist state constraining private role in policy making and in defining social ends rather than means (Birckman 1981: 489-490).

Besides deep ideological differences both superpowers adopted militarization (*the Realist Approach*) as part of their global strategy devoting huge amounts of political and economic capital to out compete each other ignoring basic needs of human security: health, food, environment and overall issues related to poverty. It could be safely said that the Cold War period {1945 — 1990s) was a period of wasted resources and opportunities. By 1990s it was, however, becoming clear that this mutual assured destruction (MAD) was a folly and a new approach to solve global problems is needed. The changing economic and political environment in post- Afghanistan Soviet Russia, and tremendous public opinion pressure in the United States and elsewhere, required better management of resources to eradicate poverty, improve environment and health especially in Afro-Asia where most of the world's population lived. Both Superpowers realized that their prosperity and global prosperity needed a different approach to placate "too many millions whose dreams are more vulgar, more real, and whose raw energies and desires will overwhelm the visions of elites, remaking the future into something frighteningly new" (Kaplan 2000: 48) —a more 'holistic', just and a more balanced approach to social, economic and political issues — ethnic, religious conflicts, looming Islamic fundamentalism, nuclear proliferation, terrorism, political instability and environmental degradation. The fall of the Berlin Wall in 1989 and unrest in other parts of Soviet occupied Eastern Europe, and the disbanding of the Warsaw Pact in April 1991 for all practical purposes saw the demise of the Soviet Union — December 25, 1991. A euphoric Fukuyama observed that with the collapse of the Soviet Union and Communism, "What we may be witnessing is the end points of mankind's ideological evaluation and the universalization of Western liberal democracy as the final forms of human government" (1992: 19). These events of 1989-91 overturned an international power structure that had dominated since WWII. It was observed that the Marxist-Leninist economic and political system had failed. Huntington spoke of a 'clash of cultures' and new global political and economic alignments, Fukuyama predicted 'the end of history' while Kaplan worried about the 'coming of anarchy'. Among this intellectual babble a search ensued for a model which would support an American (Western) led vision of liberal democracy and free market capitalism defined by the so-called 'Washington Consensus', the intellectual godfather of globalization. It was pointed out that free flow of goods' ideas, and technologies across national borders with minimum state intervention/regulation will create a global system of economic prosperity, democracy, and international peace and harmony. Since democracies and countries which trade together do not go to war, the new world order will rely on soft power. Globalization became the new mantra, the second coming of Christ, and the buzzword of the 1990s. These differences in expected benefits aroused several issues such as:

- The growth of transnational/multinational corporations and the spread of global capitalism
- Weakening of the sovereign states due to the enhanced role of existing global governance bodies (Vajpeyi et. al 2015)

DEFINING GLOBALIZATION

Since globalization means different things to different people, and is quite a complex concept, the "great globalization debate" is between those who are skeptical about the process (*The Skeptics*) and those who accept it as a reality (*The Globalists*) (Held & McGrew 1999). Following is a brief summary of main elements/aspects of globalization (Albrow 1990; Cox 1993; Khor 1995; Held et. al 1999; Rothkopf 1997; Akoi 2002; Webb 2006; Strange 1996; Stiglitz 2002).

- Internationalization of finance
- Increased international trade
- Adoption and application of new technologies
- Reduced role of the sovereign state, and the emergence of non-state actors such as non-governmental organizations (NGOs), an increased role of multilateral agencies (the World Bank, the IMF, the WTO, etc.), multinational corporations (MNCs) impact on global governance
- Increased transborder migration
- Homogenization of cultures
- Spread of democratic values

An appraisal of the impact of globalization provides us mixed results (Vajpeyi et.al.2018). Following is a brief discussion of the impact of globalization on several aspects of social, economic, and political facets.

Globalization, Democratization, and Social Capital

Proponents of democratization from Lipset (1960) to Milton Friedman (1962), the World Bank, International Monetary Funders, Bhagwatis (2002), Gilpin (2000), (Kaplan 2000: 23) and many others have consistently maintained that free market at global level accomplishes three things:

- Weakens the coercive power of the state
- Creates a more democratic middle class
- Transforms political culture by exposing masses to democratic-participant-liberal ideas.

Opponents of above arguments maintain that capitalist transformations weaken rather than strengthen liberalism (Moore 1993). For example, multinational corporations in the name of

Globalization demand greater tax and investment freedoms and privileges and put business first over democracy.

As discussed above, globalization's impact on democratization generates forces which reshape social capital and help in the emergence of a civil society as a significant political force (Tulchin et al 2002:116). Social capital is "Civic engagement involving both private and political organizations." (Mauceri 2008: 10) Foley and Edwards (1996: 38-42) maintain that "social capital is the glue of civil society helping to determine the density and bonding of a broad range of social organizations and relationships," however globalization's impact varies and has different consequences for different societies.

Table 2: Impact of Globalization on Social Capital and Implications for Democracy

Globalization Processes	*Social Capital Outcome Scenarios*	*Democratic Challenges of Social Capital Outcomes*
Market Liberalization	1) Decline and Atomization	Collective Action Problems
Communication Revolution		Demagoguery/Opportunism
Consumerism	2) Issue/Identity Transformation	Disjunction between issues and institutions: unresponsiveness
Limitations on the State		Exclusion/Conformity

(Mauceri 2008: 10)

Weakening of the Sovereign State and the Role of Governance

It has been observed that globalization in the name of integration not only seeks a new configuration of the international system but it also tends to influence the traditional role of the Westphalian sovereign state. Traditionally the state through its regulative, distributive, and coercive powers consolidates different groups in the society. It acts as a judge, jury, and executioner. It determines the economic and social policies to maintain a working balance between competing groups in society. Globalization by its very nature intrudes into vital spheres of national policies and thereby sets itself as 'competitor' with a great degree of autonomy and freedom. Globalization is seen "as an ever tightening vise increasing the demands on states to provide safety nets while limiting their ability to do so" (Annan 2000: 126j. Also, as "globalization proceeds, state roles and structures experience significant transformations. Weberian bureaucracy is becoming New Public Management (NPM), while the Keynesian model that previously formed state-economy relations is losing its hegemony to neo-liberal and monetarist theory" (Sven 2004: 282). The debate between the "proponents of state redundancy versus the champions of continuing state potency" presents the following arguments (Clarke 1990: 44-45).

1. Since national economies will be part of the global economy the role of the state in defining and managing economic policies will be reduced (Reich 1991: 3; Higgat 1996: 33; Rosenau 1997: 362). George Soros (2002) an ardent supporter of globalization concluded that "international finance and trade have outstripped the capacity of sovereign states to manage the politics of globalization."
2. Success of free open market global economic system depends on free flow of goods and finance across borders. It invites powerful transnational corporations (TNCs) to heavily invest in countries such as China, Mexico, and Thailand, which provide them a hospitable business climate — weak or non-existent labor unions, lax environmental laws, and large tax benefits. In many cases these TNCs have more financial resources than the total annual budget of the host states. Also most of these TNCs are owned by Western/European conglomerates. They receive strong political support from their respective governments. It has also been pointed out that these TNCs invest very little, if any, in delivering and supporting social policies helpful to the weakest in the society, those who might have lost their farmland, or jobs due to the transnationalization of native businesses. In fact they exacerbate and even exploit the vast social and economic inequalities that exist in the world. These developments encroach on the state's power by subverting the state's capacity "to control and protect the international life of society" (Falk 1997: 124-25j. It is also expected that in a globalized system trading partners will become political partners, and the economic interdependence would eliminate the potential for political and military conflicts (Annan 2002: 127). It has turned out to be a wishful thinking and a false promise. As discussed earlier, globalization has neither brought economic prosperity to most of the people in the world nor has it ushered democracy especially in those countries which seem to have benefited most from global investment. In many countries — Asia, Africa, and Latin America — the gross domestic product has declined, and especially neglected have been public health, human

rights, environmental protection, and other public goods. Those at the bottom are not doing so well at all. The states' capacity to manage these issues has not increased with globalization. A diminished state cannot provide adequate safety nets. There have been more losers in these economies than the winners.

3. Above discussion pointed out the negative impact of globalization on the sovereign states' ability to make policies in their own borders; however, there are those who are not ready to bury the state. They maintain that globalization does not weaken the states' power. It expands and strengthens the role of the state in an international framework of consensus and compromise based decision making. A decision by the WTO is as much binding on the U.S.A. as it is on Botswana. "This side of the argument is expressed once again in three principal versions:

- That all economic and market structures reflect political frameworks and choices;
- That the state remains potent in the development of the process of globalization itself; and
- That the state remains decisive, notwithstanding the effects and constraints of globalization" (Clarke 1999: 45).

In my view traditional concept of state sovereignty has been going through change and transformation for a long period of time. With increasing interdependency, the role of the United Nations and many other multilateral agencies, the World Bank, IMF, the WTO, emerging international NGOs, and concerns about environmental issues have already introduced subtle and not so subtle changes in the behavior of states around the world. Very few countries are willing to ignore international public opinion on many global issues related to human rights, terrorism, environmental protection, and so on. In the 21" century the state will be trimmer, relatively more transparent irrespective of purely economically driven globalization. As long as it is realized that "no one wants to reside in a totally closed society like North Korea, but complete integration and openness also bring their perils and achieving a fine balance between accessibility and security will be excruciatingly difficult" (Kennedy 2001: 43).

Washington Consensus policies were based on a rejection of the state's activist role and on the promotion of a minimalist, non-interventionist state. The premise was: governments are worse than markets, the smaller the state the better the state (Stiglitz 1998a: 25). Washington Consensus-inspired reform policies were market-biased. They were unable to keep the adequate balance between state and market and often disregarded the importance of improving the public sector, since they believed in the market's capacity to solve all basic problems of society (”market fundamentalism”). This belief also led them to ignore the role of the state in a sound economy and to support trickle-down theory in spite of its failure (Stiglitz 2003: 21 and 27).

Stiglitz criticizes Washington Consensus' view on the basis of theoretical and empirical foundations.

Market-failure theory provides the arguments for an economic justification of state intervention. Market failure happens when the market is unable to allocate resources efficiently *in Pareto's terms*. Six kinds of market failure have been identified: lack of competition, public goods, externalities, incomplete markets, imperfect information and unemployment. Even if market failures were amended through state intervention and *a Pareto-efficient* resource-allocation were achieved, it might be a very unequal one. So income redistribution is another task the state should undertake (Stiglitz 1992: 74-85).

This alternative perspective regards the government (the state) as a catalyst and as a complement to markets (Stiglitz 1998a: 25). The "question should not be whether a particular activity should be carried on in the public or private sector, but how the two can best complement each other, acting as partners in the development effort" (Stiglitz 1998b: 25). Government as a complement should both (1) provide resources to the market — which, if left to itself, tends to systematically underprovide in arenas such as human capital (public education, access to funding) and technology (investment in the production and adoption of new technologies) (Stiglitz 1998a: 26-27); and (2) create an enabling environment for the private sector and fight poverty (Stiglitz 1998b: 26). According to Orlansky et. al (2008: 200-201) the Asian miracle in general, and China in particular, is the decisive event for Washington Consensus' critics when considering the issue of government intervention. Asian countries have shown that a successful development process, involving both poverty reduction and widespread improvements in living standards, could be achieved with the government playing a large role (Stiglitz, 1998b: 10).

B) Global Governance

Quite often the British poet, Alexander Pope, has been quoted that "for forms of government, let fools contest, whatever is best, administered is best," has validity but institutions which provide a framework do matter North 1989 analyses the new classical theoretical framework by stressing the pivotal role of institutions in economic governance. Politics and annunciation of polity matter but nevertheless it is the operationalization of these policies in the form of good governance local state and global this is key to improving the quality of life secure in human security and attaining sustainable development.

According to German philosopher Hegel, define the crucial relationship between civil society and the state. Governance enables "participation of those concerned in the formulation and development of policies and programmes and practices for equitable and sustainable system" (Tulchin 2002-3) in a globalized democratic contact institutional capacity building macro level and strengthening of citizen participation (micro-level) are essential for good governance. "Those who govern are responsive to the wishes and demands of citizens and various groups formulation, adoption, execution, monitoring, and evaluation of governmental programs... which require accountability of those who govern their constituents within a system of laws, rules, regulations, and Standards; mobilization of resources to pursue the developmental vision and achieve good outcomes" (LaQuian 2002: 99). In an ideal world if all requisites of good governance could be achieved a globalized world would be politically, economically, and socially acceptable system.

It is hoped that globalization will contribute to 'Internationalization of the state' and 'of societies' (Goldman 2001; Friedman 2005) and will increase the international Treaty of political decision making (Goldman 2001) which will need the harvesting of technologies in general and information Technologies (IT) in particular.

Genesis of Global Governance

In pre-WWII Global governance was simply Colonial rule for most of Asia, Africa and Latin America. Policies were made in Paris London Lisbon and Madrid and executed by representatives of these powers with some help from native administrative staff. There was hardly any input/concern about the subject basis. George Orwell's 'Burmese days' is a very interesting (besides many others) description of colonial administration. The End of World War II brought many crucial economic and political changes. The most prominent was the restructuring of global power systems (1) the emergence of two super power blocs America led Western Europe and the Soviet led Eastern Europe and the (2) end of colonialism. The war not only rooted forces but it also totally devoted a large part of middle and Eastern Europe dash its economy and societal infrastructure full stop there were a few muted suggestions to reduce Germany to its pre-industrial page. More pragmatic heads prevail American policymakers decided to rebuild your of- economically and politically and socially by introducing a very ambitious state and nation building Marshall Plan. It had two broad goals; (1) not to repeat mistakes made by Victoria Ellis after World War first by humiliating the vanquished Germany. (2) Reconstruct European society is to prevent future conflicts. Related to these goals was a very strategic Long Term plan not only to establish itself America as a human but also to contain threats from Marxist communist Soviet Russia by bringing these Western European countries under the military alliance – the NATO full stop to meet these goals and other challenges new Institutions of governance were needed. A weak League of Nations was long dead.

Bretton Woods Institutions

The allied States the US and its Western European allies created a triad of global Institutions to stabilize the global economy. In 1944 forty four allied States met in a conference at Bretton Woods New Hampshire (USA) to finalize the World Bank's articles of agreement was stopped by late 1945 these articles were ratified and thus were created an International Bank of for reconstruction and development IBRD/the World Bank. As mentioned earlier the primary purpose of earlier World Bank project was the reconstruction of shattered economies of western Europe however it was realized that there were serious problems of poverty malnutrition literacy at which played recently independent former colonies in Asia and Africa and these might spillover in other parts of the globe endangering peace and prosperity to stop the World bank therefore redirected its mission and became involved in economic development projects in these areas.

It was almost an accepted view among economists that development occurs most predictably and readily when it rises in a context of appropriate government policy. The World Bank and its sister institution – 'The Bretton wood Twins' – the International Monetary Fund (IMF) do not agree about the contacts and substance of appropriate government policy. funding development projects in poor countries remains the world Bank's primary activity while the IMF basically monitors and imposes conditionalities and a link economies and has been quite an popular for its unreasonably hard behavior several other affiliates of World Bank such as International Finance Corporation 1960 and regional banks in Asia Africa and Latin America also fund development projects.

No doubt that Bretton Woods Institutions contribution to global economic development has been admirable over the years but they have also been criticized for being very aggressive agents of Western models of capitalism and global governance and oftentimes even promoters of American foreign policy.

Institutions matter but…

As discussed earlier that decolonization brought political independence but also the challenges of state and nation building. The creation of worldwide institutions provided new framework for early Global governance. Unfortunately these institutions and their work was seriously impeded by cold war rivalries, corruption and inefficient administration. Besides these difficulties Global political and Economic map was changing very fast. Globalization despite its many shortcomings was having an impact on global economies and global power structure. New actors were emerging- BRICs- new challenges and opportunities created by fast changing technologies required different versions and agencies to fulfill them. 'Globalization and its Discontent' challenges status quo pointing out flaws in existing global governance bodies. These flaws include (1) structural issues, (2) policy making processes, (3) adherence to rules, (4) participation (Koppell, 2010: 271)

Democratic Deficit

One of the most serious and persistent criticisms of three major governance actors the UN the World Bank and IMF- has been their 'non-representative' nature of membership. The United Nations Security Council has only one non-western member (China). While Europe/ Eurasia (Russia) has three members plus the United States, Latin America, the Middle East and Africa do not have a single member. The Financial Stability Forum includes the Netherlands and Australia, but not China or India, the two fastest growing Global economies. At the World Bank, and the IMF, Africa has very few seats. "The IMF", it has been pointed out, has a, "penchant for secrecy," (Stiglitz 2000: 228). The UN has also been criticized for incompetence and fecklessness (Bayefsky 2007; Hawkins et. al 2006) and even dismissed as irrelevant (Morgenthau and Thompson 1993; Waltz 1993; Gilpin 2002). Martha Nausbacum denounced the United Nations' system as "grotesquely flawed and corrupt, totally lacking in democratic accountability and therefore devoid of any procedural legitimacy when it comes to imposing law on people and good governance," (Mishra 2018). Even in the most crucial and sensitive issue on human rights United Nations shows several faces. It supports a so called "Universal system in the service/promotion of healthy global/regional communities, a religiocentric model of Judeo-Christian value system. Universal value system must not present Ethnocentric, Eurocentric models and expect other (non-Judeo-Christian) cultures to embrace them without any say. Authoritarian disrespect for truth or for democratic institutions, and little comprehension of or concern for the public good" (Freeman 2017: 65) defeats any effort to construct a healthy community- global or otherwise and poisons global governance If the World Trade Organisation (WTO) negotiations on trade agreements are done in secret making it difficult to see the influence of major multinational corporations (MNC) and other vested interests. (Stiglitz 2002: 227)

Lack of Accountability and Legitimacy

International organizations such as the IMF and the World Bank have been faulted for their lack of accountability and legitimacy and quite often as tools of western economic policies. "Accountability is a matter of fundamental importance. It is a gateway to core matters, including the very legitimacy of the organization in question...overcoming such legitimacy problems is necessary for global governance organizations," such as the World Bank, the IMF and the WTO (Koppell 2010: 27). Bakari (2018: 119) points out that under present global governance administered by major international institutions a conflict occurs between sustainable development and agendas of these Global actors. "Whenever economic growth clashes environmental concerns, economic growth almost always wins." (Robins 2009-12) as a result "global economic and political governance…sidelines ecological and economic considerations." (Kutting 2004:3) In the operation of these major global governance agencies detects one common element of inequality and unbalanced influence, however it should be accepted that some progress has been made to satisfy the critics due to strong leadership of China and India, and some concessions/accommodations made by status quo powers especially Europeans and Japanese, yet much is desired.

What is to be done?

The post-Cold War Era has witnessed several monumental changes in international economic and political Spears. The collapse of the Soviet Union in 1991 left the United States of America as the sole superpower in the Global Arena with unprecedented economic and military power and tremendous freedom of action which led to the "Unilateral impulse in American foreign policy," (Gilpin 2002) the 'arrogance of power' which often meant treating others as 'instruments in a concert with an American conductor," (Hoffman 2000: 6). Aggressive global efforts by American policy establishment invoked the free market visions of Hayek and Adam Smith to solve the global economic problems within an American defined- Washington consensus- globalization, and democratic values. As discussed earlier that global free trade did lift hundreds of millions of people in China, India, and elsewhere from poverty, strengthened multinational corporations, questioned the role of nation state in economic development but unfettered, unregulated market created wealth inequality and environmental degradation and fundamentalist ethnic impulses. It turned out to be a false God and by 2008 it started to lose its shine. The failure of Lehman brothers on September 15, 2008, the housing bubble, and high unemployment raised serious questions/doubts about the viability of the Washington Consensus based globalization and to existing global governance agencies ability to cope with these problems. Even Greenspan, the former US chairman of the Federal monetary policy making body, once known the "Maestro of capitalism" declared himself 'in a state of shocked disbelief'. Prominent Scholars and policy makers who were vociferous in supporting the globalization model started to express their serious doubts, and talked about finding a new model- an alternative model(s). It was pointed out that both China and India had weathered the recession storm better with less dislocation then the United States and other Western democracies. Both China and India have adopted certain practices of free market economy but had kept state as the main guarantor of national interest related economic sectors. Their approach came to be known as State capitalism. In my view the advocacy of unabashed and unregulated Washington consensus driven globalization is 'Capitalist narcissism' denial of other ideas/models. 2008 financial crisis shattered this obsession with this Ideological virginity. Let us very briefly discuss the frequently suggested native models of global governance and economic development (Table 3)

Table 3: Native Models of Global Governance and Economic Development

China Model	*India Model*
Highly regulated National interest sectors as (SOE); relaxed open free market export terms, 'hard authoritarian political system' (Shambaugh 2016) 'Adaptive authoritarianism'	Regulated national interest sector, limited foreign investment-export oriented area, relaxed, democratic multi-party political system

Source: Vajpeyi (2017)

The India Model

From its very inception, the Indian movement for independence had both political and economic agendas. Ideas of economic equality were an integral part of political freedom and equality. The economic ideologies of various political groups which had joined the umbrella organization – the Indian National Congress - during the Independence movement favored a socialist democratic India with emphasis on land reforms, industrialization and economic growth. They believed that the state had a prominent role in nation building and in distributive process by providing an efficient administrative system at home, and also by actually participating in global governance. Post-independence Indian governing elites were convinced that planned highly regulated economy under five year plans will establish a democratic socialist India providing opportunities to all sections of Indian society to unchain themselves from poverty. However, to the dismay of Indian elites and the general populace five year plans failed to deliver on their promises. Political and administrative corruption and inefficiency both at national and state levels thwarted the progress. High regulations gave unprecedented discretionary powers to bureaucrats. India came to be known as 'Permit Raj' (governance by regulations) and was ridiculed for its "Hindu rate of economic growth", - slow and stagnant. The Nehru model of economic growth was deemed a failure. By 1991 India's economy went into a "tailspin after its foreign exchange reserve dropped almost to zero in the aftermath of Gulf War," (Luce 2007: 30), and "Nehru's Socialist dream of creating an economy that would be immune from influences of the former colonial powers had culminated in bankruptcy." (Luce 2007: 30-31). Drastic measures and embarrassing compromises were required to avoid total economic Chaos and potential political instability. Ironically, these new measures to open India to foreign investments were engineered by Nehru's grandson Rajiv Gandhi. India joined the global economy, and since 2014, its economy had been moving faster. "Modis has lowered barriers to foreign investments and dramatically simplified the tax system. India's economy has taken off with growth forecast at 7.8%," (Bremmer 2016: 8). India's economy has an average expanded at 6% year since 1991, roughly doubled the "Hindu rate and measured by quality, if not quantity," (Luce 2007:49). It must be noted here that unlike China, India functions in a very chaotic noisy political environment. Even simple decisions required deft political skills to satisfy competing regional and sectarian economic and political interests under a very independent judicial system. India is not out of the woods and its economy faces many domestic and external hurdles and challenges. It has miles and miles to go especially in a very unstable International environment being created by conclusions among EU countries and above all by a whimsical President of United States Mr. Trump. "Fiery control stations between backers and critics of the BJP", globalization on University campuses are quite frequent appearances while Modi has accomplished a lot and India remains mostly a positive story. But the pace of progress is about to slow and the ghoots from the past are now circling overhead." (Bremme 2016:8)

Especially if left leaning almost anti-Western Congress party games in future. The Indian model's aim is "may all be happy, may all be free from illness and disease, May all be safe and secure". It must be noted here that is fired several suggestions by economists and policymakers most Americans- India has not touted its old model of governance. China presents a very different story.

The China model: (Crossing the river by feeling the stones)

One of the global changes of historic significance took place in 1948 when after a long civil war China was liberated from its feudal shackles by the Chinese Communist Party (CCP) under the leadership of Chairman Mao Zedong. For decades the Chinese Regime was not 'recognized' by Western powers led by the United States of America and was denied its rightful place in international organizations such as the United Nations, but ultimately political realities of changing strategy interests forced China to become full-fledged member of the global community, including membership in the United Nations Security Council. Unburdened by past restraints, China launched ambitious, massive programs to modernize its economy. In December 1978, at the Third Plenum of the Eleventh Central Committee of the CCP, Deng Xiaoping presented a blueprint of 'Four Modernizations' for the transformation of Chinese economy. Past seventy years of efforts and dedication by three major modernizers (1948-2018), Mao, Deng and Xi have reshaped China. China is not a developing country (in a traditional sense), anymore. Its economy is second in the world and its Global presence is felt on issues as diverse as the environment, free trade and global governance. Chairman Mao romanticized revolution, and as president Xi Jinping has credited Mao with "creating fundamental political conditions and the institutional foundation for achieving all development programs in China today", and Deng Xiaoping with, "launching the new great revolution of reform and opening up", President Xi announced the arrival of a "New Era" with a goal to create a "moderately prosperous society", a "modernized socialist society" by the hundredth anniversary of the creation of the Chinese Communist Party (CCP)" China presents the tale of three modernizes. But Xi's dream of rejuvenating China differs from those of his predecessors in his passion for order, a value shared neither by Mao who loved chaos, nor by Deng who loved to experiment. Xi unpacked the contradiction in Mao's thought between Chaos and conformity and has chosen conformity; he has rejected Deng's acceptance of the fact that human nature is unalterably selfish but has embraced Deng's faith that prosperity will create stability." (Nathan 2018: 36). How does president Xi achieve his objective which could present a model of governance both for China and abroad? It has been found out that

- Xi is making the Chinese institutional system (political and administrative) he inherited more responsive and efficient. Eradication of corruption has been his priority. It is reported that under his leadership more than a million party members including 170 at higher levels have been disciplined. (Nathan 2018: 37)
- Call for the merger of core socialist values and find traditional Chinese culture. Today "China presents itself as a society with an extremely successful capitalist system. Part of China's claim for World leadership does increasing rests on the argument that China with its Confucian form of governance will be able to tackle issues like economic inequality and climate change- issues that the United States, with its governance structures so controlled by lobbying

> interests, cannot address." Confucian thought is "now being presented as embodying a vision that offers new possibilities for 21st century humanity." (Puett 2018: 234) "an alternative to the alienation, individualism and anthropocentrism of Western modernity in general and more specifically to the dysfunction of forms of governance seen in western societies." - Soulless materialism of the west. How this will work out in modern China is yet to be seen.

Now let us turn to recent discussions taking place in China itself about the China model and its relevance to global governance. In a Chinese Publication "Future of Global Governance China Perspective" (Beijing: SPIGG 2016) it is observed that Western "systems, democracy and model of economy have failed to show superiority. Given the insufficient representativeness of and effectiveness of current global governance framework the world is expecting China, a representative of the emerging countries, to offer its suggestions and proposals for advancing global governance reform… to build a community of a shared future "for mankind…a new concept and philosophy put forth by the Chinese government. (SPIGG 2016: 34). While this statement does not directly propose the China model as enunciated in the Report (34-36) to be adopted by other developing countries but by claiming to be the representative of developing countries it is expected to adopt suggestions and leaders of China. The report further presents Chinese perspective for global governance. Following is a brief summary of Chinese perspective on global governance:

- To improve the mechanism of financial stability board, speed up the formulation of norms for International Financial regulation
- To set up a G20 secretariat to coordinate the communication within the G20
- To build Silk Road Economic Belt and the 21st century Maritime Silk Road in which China and Asia will play a prominent role
- To reform decision making process of IMF, the World Bank, WTO and increase the representation of emerging economies
- To accede to the Paris Agreement on Climate Change

And last but not least global governance/good governance cannot be achieved by threatening other sovereign nations, pulling out of already accepted international agreements, and imposing tariffs on the fake and dubious 'national security' excuses. Tweets are for twits. International/Global Cooperation can be achieved only through respectful negotiations among community of nations.

This above discussed alternative models have their attractive sides and achievements but they also have like Washington Consensus based globalization their limitations. Ruchir Sharma's "research suggests that compared with democracy is autocracies generate far more unstable growth on a list of thirty six countries that have been whipsawed growth and recession throughout the post-war era, three out of four autocracies" also Sharma refutes the argument that state capitalism model provides stability as well as growth (2016). The "State capitalism's biggest failure is to do with Liberty. By turning companies into the organs of government, state capitalism simultaneously concentrates power and corrupts it. It introduces commercial criteria into political decisions, and political decisions commercial ones. And it removes an essential layer of scrutiny from central government. (Wooldridge 2012-18). The present neoliberal globalization has its flaws as evidenced during the 2008 recession. It requires adjustments and less ideological hubris from America but we should not try to stop globalization even if we could. But if we don't do a better job managing a changing world economy it seems clear that it will and badly The New York Times may 26 2016). A world of nation fortresses will be poorer and gloomier (Wooldridge 2012 22). To quote Karl Marx (from his grave-stone) "the philosophers have only interpreted the world in various ways; the point however, is to change it." Blaming globalization and abandoning its core values will be a retreat to a world of no rules and parameters.

President Obama said that globalization is an Irreversible reality. "The integration of national economy into a global economy that's here. That's done. So the question is not whether or not there is going to be an international global economy. There is one, "at a meeting of North American leaders in Ottawa in June 2016 Obama emphasized that "The question is under what terms are we going to shape that economy…And us trying to abandon the field and pull up the drawbridge around us, is going to be bad for us." (The Weekly Standard, Obama says globalization is 'Here' and 'Done' 29 June 2016). Let us note that events move history and if…" taken at the floods, leads on to fortune; omitted, all the voice of their life, is found in Shallow and miseries"- Shakespeare.

How will capitalism end/ will it?

More events in Europe and elsewhere- immigration crisis Brexit rise of Alt Right movements in France, Hungary, Poland, Austria and the election of a populist 'destabilizer' Trump as President of The United States of (the home of 'Washington Consensus') and his vocal support to economic and political nationalism (American First) have started vigorous debates about the future of the capitalism itself (Ash 2017: 19-24;Wolfgang 2016; Bromwich 2017; Priestland 2017). "What makes capitalism toxic is its expensiveness". Capitalism destroys its own foundations. It undermines the family units on which the reproduction of labour depends. It consumes nature, it commodifies money… For its own good capitalism needs political checks," (Tooze 2017: 3). But despite differences of opinion doubts and worries about the end of capitalism, most of these discussions do not see the demise of capitalism, and do not envisage the return of Marx and the withering of the State. (Stedman 2016; Menand 2016:90-97). The general consensus accepts inherent weakness of unbridled, extremely unregulated free market driven capitalism. "Bringing capitalism back into the ambit of democratic government, and thereby saving the later (capitalism) from extinction means de-globalizing capitalism (streeck 2016) "there might be challenges Liberal order but they would come you from Islamic aur China state capitalism no longer a discredited Marxism. (Priestland 2017). Capitalism is not ready as yet to be doomed in "dustbin of history". Habermas "foresees the possibility of a positive transformation of capitalist society into a democratic Society of 'domination-free communication' where members are aware of and can publicly acknowledge and accept the basis of these social relations and democratically decide their own fate. These conditions….are essential to overcoming the illness caused by ideology" (Freeman March 23/ 2017:65). "Unlike the neo-Marxists…Habermas rejects communism and advocates a liberal democratic constitution, a market economy and a local democratic welfare state that protects workers' rights of representation and codetermination in corporate decisions" (Freeman, March, 23/2017: 65). Habermas does not reject 'democratic liberalism' which constitutes the very core of capitalism. Capitalism is not dead and will continue to influence the "Human creativity, spontaneity and freedom of the self". It is part of human nature that "In periods of economics stress all sorts of theories are entertained about the nature of the problem. When better times return, some theories fade from memory. Others linger, however. During the economic mess of the past decade, economists frightened themselves with tales of 'secular stagnation' a nasty condition that dooms its growth. Now that the economic outlook is brightening a bit- deflation has been dispatched, and for most advanced economies 2017 is forecast to bring a third consecutive year of economic growth- it is tempting to laugh of the idea of 'secular stagnation' as a bit of crisis-induced hysteria. Tempting but also premature.'(The Economist, March 11/2017: 14)

REFERENCES

1. Akoi, T (2002) Aspects of Globalization in Contemporary Japan, in Berger, P.L and S. Huntington *'Many Globalizations: Cultural Diversity in the Contemporary World'*, New York: Oxford University Press.
2. Albrow M. (eds.) (1990) *Globalization, Knowledge and Society*, London: Sage.
3. Annan, Kofi (2001) '*IT Challenge to Silicon Valley*' www.news.com/ Kofi Annan.
4. Bakari, Mohamed-el Kamal (2017) *The Dilemma of Sustainability in the Age of Globalization*, London: Lexington.
5. Bayefsky, Ann F (2007) *The UN and Beyond: United Democratic Nation*, New York: Hudson Institute.
6. Bhagwati, Jagdish (2002) *Free Trade Today*, N.J. Princeton University Press.
7. Bremmer, Ian (2016) *India's Economic Engine has a Rocky Road Ahead*, Time, September 5.
8. Brickman, R. (1981) *The comparative Analysis of Science and Technology* (Review article) Comparative Politics July.
9. Broomich, David (2017) *Act One, Scene One*, London Review of Books, February.
10. Chan Koon Chung (2009)*The Fat Years,* New York: Anchor Books.
11. Chasek, P et. al (2010) *Global Environmental Politics Dilemmas in World Politics*, Colorado: West View.
12. Clarke I (1999) *Globalization and International Relations Theory*, U.K. Oxford University Press.
13. Cox, R.W. (1993) Production and Security in David B. Dewitt et. al. (eds) *Building a New Global Order*, Toronto: Oxford University Press.
14. Dwivedi, O.P. (2013) *Perspectives on Governance and Society*, Jaipur: Rawat Publications.
15. Falk, Richard (1997) *State of Seige: Will Globalization win out*: International Affairs, vol. 73, no.1.
16. Finkelstein, Lawrence (1995) '*What is Global Governance*', Global Governance 1(3): 367-372.
17. Foley, M. and B. Edwards (1996) *The Paradox of Civil Society*, Journal of Democracy, July.
18. Freeman S. (2017) *The Headquarters of Neo-Marxism*, the New York Review of Books March 23.
19. Friedman Milton (1962) *Capitalism and Freedom*, Chicago: University of Chicago Press.

20. Friedman, Thomas (2005) *The World is Flat*, New York: Ferrar, Straws and Giroux.
21. Fukuyama F. (1992) *The End of History and the Last Man*, New York: Free Press.
22. Gilpin, R. (2000) *The Challenge of Global Capitalism the World Economy in the 21" Century*, Princeton, N.J. Princeton University Press.
23. Goldman, It. (2001) *Transforming the European nation-state: Dynamics of Internationalization*, London: Sage.
24. Hawkins, Darren G et. al (2006) *Delegation and Agency in International Organizations*, New York: Cambridge University Press.
25. Held, D. and McGrew A (1999) *Global Transformations, Politics, Economics and Culture*, Cambridge: Polity.
26. Higgat, R. (1996) *Beyond Embedded Liberalism: Governing International Trade in an Era of Economic Nationalism*, in Gummet P (ed.) Globalization and Public Policy, Cheltenham, U.K.: Brookfield.
27. Hoffmann, Stanley (2000) '*Too Proud'*, The New Republic.
28. Jain, R.B. (2005) '*Globalization and Good Governance Pressures for Constructive Reforms*, New Delhi, Deep and Deep.
29. Kaplan R. (2000) *'The Coming Anarchy': Shattering the Dreams of Post-Cold War'* New York: Random House.
30. Kennedy P (2001) *A Herculean Task: The Myth and Reality of Arab Terrorism* in Tabolt S. and Chanda N (eds) The Age of Terror: America and the World After September 11 New York: Basic Books.
31. Khor, M 1995 '*Development, Trade and The Environment: A Third World Perspective* in Goldsmith E. et al (eds) 1995 The Future of Progress Reflections on Environment and Development Devon: Green Books.
32. Koppell, Jonathan (2010) *World Rule: The Politics of Global Governance;* Chicago University of Chicago Press.
33. Kutting, G (2004) *Globalization and the Environment Greening Political Economy* New York New York State University Press.
34. Laquian, A. A. (2002) *Urban Governance: Some Lessons Learnt* in Tulchin et al (2002).
35. Lipset, S.M. (1960) *Political Man: The Social Bases of Politics* New York Doubleday.
36. Luce, Edward (2007) *In Spite Of Gods: The Strange Rise of Modern India* New York Doubleday.
37. Mauceri, Philip (2008) *Globalization, Social Capital and Democracy: The Andean Region in the Comparative Perspective* in Vajpayee and Khator (eds) 2008.
38. Mishra, Pankaj (2018) *The Make It Wears,* London Review of Books, June 20 19-13.

39. Moore, B. (1993) *Social Origins of Dictatorship and Democracy,* Boston, Beakon Press.
40. Morganthau, Hans, and Kenneth W. Thompson (1993) Politics Among Nations: The Struggle Power And Peace, New York, McGraw-Hill.
41. Murphy Craig (2002) *The Historical Processes of Establishing Institutions of Global Governance and the Nature of Global Polity in Towards a Global Polity* (eds) M.O.R. Higgot New York, Routledge, Quoated by Koppell (2010-20).
42. Nathan, Andrew (2018) *China: Back to the Future,* The New York Review of Books; May 10: 36-37.
43. North, D (1989) *Institutions, Combi, Institutional Y Decempeno Economica Magico Fondo De Cultura Economica.*
44. Orlansky Dora et. al (2008) *Does Democratization work in Argentina? Human development and governance* (1989-2005) in Vajpayee and Khator (eds) 2008.
45. Priestland, David (2017) *A Long Way From The Winter Palace,* The New York Times.
46. Puett, Michael (2018) *Who Is Confucius In Today's China, in* The China Questions: Critical Insights Into A Rising Power, (eds) Jennifer Rudolph and Michael Szonyl, Cambridge; Massachusetts: Harvard University Press.
47. Ruthkopf, D. (1997) *In Praise Of Cultural Imperialism* in O'Meara et al (eds) Globalisation And The Challenge of A New Century: A Reader Bloomington Indiana: Indiana University Press.
48. Reich, R.B. (1991) *The Work of Nations Preparing For 21st Century Capitalism,* New York and Schuster.
49. Reppy, Judith (1979) *The Control Of Technology Through Regulation* in Johnson, Ron et al (eds) Directing Technology, New York St. Martin.
50. Robbins, R. H. (2009) *Introduction Globalisation And The Environment: A Primer* in GM Kroll et al (eds) Globalisation And Development Reader, Lanham: Atmira Press.
51. Rosenau, J.N. (1997) *Along The Domestic-Foreign Frontier: Exploring Governance In A Turbulent World* Cambridge UK Cambridge University Press.
52. Schell Orvill (2016) *Crackdown In China: Worse And Worse,* The New York Review of Books April 12, 12-16.
53. Shambaugh, David (2016) *China's future,* Cambridge, UK: Cambridge University Press
54. Soros G (2002) *George on Globalization,* New York Public Affairs March 29.
55. Stiglitz, Joseph (2002) *Globalization And Its Discontents,* New York: W. W. Norton
56. Strange, Susan (1996) *Casino Capitalism,* Oxford, UK: Blackwell.

57. Sven, B (2004) *Globalization, State Transformation and Public Security*, International Political Science Review, July.
58. Tulchin J. et al (eds) (2002) *Democratic Governance And Urban Sustainability,* Washington DC Woodrow Wilson International Center.
59. Vajpeyi, Dhirendra and R.K. Mishra (2012) *Global Governance Conflict Resolution and Sustainable Development,* Delhi: Macmillan.
60. Vajpeyi, Dhirendra et al (2015) *Corporate Social Responsibility Apartment In Emerging Economies* London Lexington.
61. Vajpayi Dhirendra and Renu Khator (eds) 2008 *Globalization Governance And Technology* New Delhi, Deep and Deep.
62. Vajpeyi, Dhirendra and Roopinder Oberoi (2018) *Globalization Reappraised; False Oracle Or A Talisman*, London: Lexington.
63. Waltz, K.W. (1993) *The Emerging Structure Of International Politics,* International Security 18(2): 44-79.
64. Webb, A. K. (2006) *The Calm before the Storm Revolutionary Pressures and Global Governance,* International Political Science Review, January.
65. Wendt, Alexander (1999) *Social Theory of International Politics,* Cambridge UK: Cambridge University Press.
66. Whitefield, Susan (2018) *Silk Slaves and to Pass Material Culture of the Silk Road,* Berkley, University of California Press.
67. Wolfgang, Streeck (2016) *How Will Capitalism End?* London Review of Books, January 5 by Adam Tooze, A General Logic of Crisis, 3-8.

INDEX

L

M

N

O

P

R

S

T

U

W

X